10th Curtin University of Technology Science and Engineering International Conference 2015 (CUTSE2015)

IOP Conference Series: Materials Science and Engineering
Volume 121

Miri, Malaysia
6 - 8 November 2015

ISBN: 978-1-5108-2277-1
ISSN: 1757-8981

Printed from e-media with permission by:

Curran Associates, Inc.
57 Morehouse Lane
Red Hook, NY 12571

Some format issues inherent in the e-media version may also appear in this print version.

Printed by Curran Associates, Inc. (2016)

Institute of Physics
Dirac House, Temple Back
Bristol BS1 6BE UK

Phone: 44 1 17 929 7481
Fax: 44 1 17 920 0979

techtracking@iop.org

Additional copies of this publication are available from:

Curran Associates, Inc.
57 Morehouse Lane
Red Hook, NY 12571 USA
Phone: 845-758-0400
Fax: 845-758-2633
Email: curran@proceedings.com
Web: www.proceedings.com

Table of contents

Volume 121

10th Curtin University Technology Science and Engineering International Conference (CUTSE2015)
6–8 November 2015, Miri, Malaysia

Accepted papers received: 8 March 2016
Published online: 7 April 2016

Preface

Papers

Foreword

A very warm welcome to all participants of the 10th Curtin University Technology, Science and Engineering (CUTSE) Conference 2015.

This annual conference dates back to 2006 when the first Curtin University of Technology Science and Engineering (CUTSE) Conference was held in Curtin University, Miri Sarawak. CUTSE Conference was initially intended for Curtin's undergraduates to participate and present their research in a conference.

This year the Department of Applied Geology is organizing the 10th CUTSE Conference with support from The Institute of Engineers Malaysia, Miri Branch and the Miri Port Authority. This year, our theme is "Science, Engineering and Technology for a Sustainable Environment".

The 10th CUTSE Conference 2015 is an excellent venue for researchers, engineers, scientists, academics, professionals from industry and students to share their research findings and initiate further collaborations in their respective fields.

On behalf of the organizing committee, I would like to thank this year's sponsors and supporters. We appreciate your support for CUTSE 2015. The CUTSE 2015 conference would not be possible without the dedicated work and efforts of the organizing committee members, staffs and students who worked tirelessly in every aspect. I thank you for your hard work and commitment towards making CUTSE 2015 a success.

Selamat Datai (Welcome) and enjoy the conference.

Associate Professor Rajat Mazumder

Organizing Chairperson,

10th CUTSE Conference 2015

IOP Conf. Series: Materials Science and Engineering **121** (2016) 011002 doi:10.1088/1757-899X/121/1/011002

CUTSE 2015 Conference Committee

Chairperson	Dr. Rajat Mazumder
Deputy Chairperson	Dr. Ramasamy Nagarajan
Secretariat	Dr. M.V. Prasanna Ms. Alicia Lyndie Rogers Ms. Dora Cecelia Indin Ms. Jacqueline Martin
Treasurer	Dr. Vincent Lee Chieng Chen Dr. Stephanie Chan
Publications and Proceedings	Dr. Ramasamy Nagarajan Dr. Dominique Dodge-Wan Dr. Andrew Lawfield Dr. Neamul Ahsan Noman Khandoker Dr. Phan Thanh Duoc Dr. Wong Kwong Soon Dr. John Lau Sie Yon Dr. Jibrail Kansedo Dr. C. Palanichamy
Programme and Logistics	Ms. Vivian Dayong Mr. Pang Po Ken Mr. Kamaroizan Mr. Jason Chia Keh Leong Ms. Shubashini A/P Krishnan Mr. A.Anand Kumar Mr. Prabakaran
Sponsorships and Publicity	Ms. Cynthia Paul Sebli Ms. Illiya Amalina Ms. Natallee Faith Kwan Dr. Sharul Sham Bin Dol Dr. Raoof Gholami Mian Poshteh Mr. Daniel Tang Kuok Ho Mr. S Veeramani Mr. Suresh Reuben
Registration	Ms. Alicia Lyndie Rogers Ms. Dora Cecelia Indin Ms. Jacqueline Martin Ms. Lydia Mujah

Peer review statement

All papers published in this volume of *IOP Conference Series: Materials Science and Engineering* have been peer reviewed through processes administered by the proceedings Editors. Reviews were conducted by expert referees to the professional and scientific standards expected of a proceedings journal published by IOP Publishing.

IOP Conf. Series: Materials Science and Engineering 121 (2016) 012001 doi:10.1088/1757-899X/121/1/012001

Synthesis and characterization of $Sm_{0.5}Ba_{0.5}MnO_{3-\delta}$ as anode materials for solid oxide fuel cells

A. M. Abdalla*, N. Radenahmad, M. S. Abu Bakar, Pg M .I. Petra, and A. K. Azad.

Universiti Brunei Darussalam, Faculty of Integrated Technologies, JalanTunku Link, Gadong BE 1410, Brunei Darussalam.

*Email: 14h8851@ubd.edu.bn, Phone: +6738626040

Abstract. The material performance is a crucial issue in the current fuel cell technology, and for this reason we present this new series of Samarium family which can be used as electrode giving a high performance in a particular application. $Sm_{0.5}Ba_{0.5}MnO_{3-\delta}$ was prepared by solid state reaction method and characterized by using X-ray diffraction (XRD), and thermogravimetric analysis (TGA). Rietveld analysis of XRD data shows that the material has an Orthorhombic crystal structure with cell parameter ***a = 3.883(1) Å b = 3.8742(5) Å c = 7.762(4) Å***, in the Pmmm space group. TGA analyses shows that the materials is going to decrease by 0.32%.The density of the materials was calculated from structural refinement and found to be 8.372 gm/cm^3.

1. Introduction

Despite the continuous development of energy resource including fossil fuels, extensive efforts still paid to different types of fuel cell technologies, which mainly depending on electrochemical reactions to get high power density [1-6], and (used in versatile applications at different scales such as; transportation, remote power sources, structure, military use, remote villages, and aerospace). Moreover global warming problem requires essential findings with reliable solutions that come with the enhancement of environmental issues. The solid oxide fuel cells (SOFCs) in the last 20 years gave a noticeable achievements and exponentially increased attention because of the various advantages such as, easy in processing and manufacturing, operates at higher temperatures, handling more convenient hydrocarbon fuels , no reforming catalysts are required, and high power density can be obtained [7,8]. Typically, working with SOFCs materials requires the full understanding of the occurred reactions, specifically, the electrochemical reactions, in addition to the full characterization of the material, microstructure starting from the microscopic level [9, 10].

Anode supported fuel cell consumes more than 95% of the total amount of the material and it is very important to obtain good ohmic resistance[11,12].Cell manufacturing techniques mainly depend on the easy processing and minimization of the interfacial reactions between the electrolyte and the electrode materials [13,14]. The state-of-the-art Ni/YSZ cermet anodes are usually made of the commercial NiO and YSZ powders, are in homogenized by milling and mechanical mixing. Many literatures reported that performance and electric properties of Ni/YSZ cermet anodes are critically dependent on both distribution of Ni and YSZ phases and the microstructure [1,15]. However, Ni particles aggregate and reduce the porosity of an anode by eliminating the triple-phase boundary (TPB) required for cell operation, thereby decreasing cell performance.ABO_3 perovskite material aimed for use in fuel cells to avoid the high temperature sintering and to exhibit high electrical conductivity at elevated temperature. This can be achieved by substitution of either A or B sites with acceptor-or donor-type cations [16]. Recent research reported that samarium oxide Sm_2O_3 is a good catalyst with high activation energies[17-19]. Doping of Ce in $SmFeO_3$ results in a noticeable stability, and good modification of the material characteristics [20]. Samarium doped materials [21] shows that the conversion of the Sm^{+3} to Sm^{+2} species can happen at same time with the oxidation of the other doped material, so the possibility of such modifications open up the way for more research to use samarium based material as an electrode in SOFCs.

Form the pervious investigations done by literature and the good properties of Sm, a new series of Sm family was prepared and investigated as anode material. New samarium based materials $Sm_{0.5}Ba_{0.5}MnO_{3-\delta}$ was prepared by solid state reaction method at 1400 ° C then was characterized by X-ray diffraction (XRD), and thermogravimetic analysis (TGA).

2. Experimental

$Sm_{0.5}Ba_{0.5}MnO_{3-\delta}$ (SBMO) was prepared by solid state reaction method. The material SBMO anode was obtained by calcinations of the single pervoskite $Sm_{0.5}Ba_{0.5}MnO_3$ in air from 800 °C to 1400 °C within 25 total sintering hours through three stages. The initial structural characterization of the SBMoxide was performed by XRD, SIEMENS Diffraktometer D5000 Kristalloflex, Cu K-Alpha1, 40 kV, 30 mA (Chalmers, Sweden). In addition to TGA Simultaneous thermal analysis - NETZSCH STA 409 PC Luxx, run under N2 in 20 ml/min (Chalmers, Sweden), was performed at 1000C°. Finally for the density measurements, it was calculated using both experimental method based on Archimedes and theoretical method from structural refinement.

3. Results and Discussion

3.1. Structural analysis

The XRD powder diffraction patterns of the $Sm_{0.5}Ba_{0.5}MnO_3$ (SBMO) material was obtained from sintered pellet as shown in Fig. 1. The rietveld analysis of XRD data in Fig.2 shown that the material has main phase of crystal structure of orthorhombic crystal structure, and fig 3 shows the 3D crystal structure and the atoms arrangement, also the cell parameter in this structure are, a = 3.883(1) Å, b =

IOP Conf. Series: Materials Science and Engineering **121** (2016) 012001 doi:10.1088/1757-899X/121/1/012001

3.8742(5) Å, c = 7.762(4) Å, $\alpha = \beta = \gamma = 90^{\circ}$, in the Pmmm space group and very small phase of hexagonal crystal structure with cell parameters a = 5.6593(3)Å, b = 4.67012(1)Å, $\alpha = \beta = 90^{\circ}$, and $\gamma = 120^{\circ}$ in the R-3m space group for the $Sm_{0.5}Ba_{0.5}MnO_3$. Sengodan et al. [13] has analyzed the layered perovskite $PrBaMn_2O_{5+d}$ and showed very good performance as anode materials. However their structure refinement was a mixture of cubic and hexagonal phases. Due to splitting of some major peaks in our diffraction pattern, we got orthorhombic and hexagonal phase which converged in a very good χ^2 value ($\chi^2 = 0.515$). This suggests that the perovskite structure is well developed after the certain calcinations process with elevated temperatures [20].Corresponding data (2θ, D_{hkl}, and h, k, l values) for this sample is presented in Tables 1–2

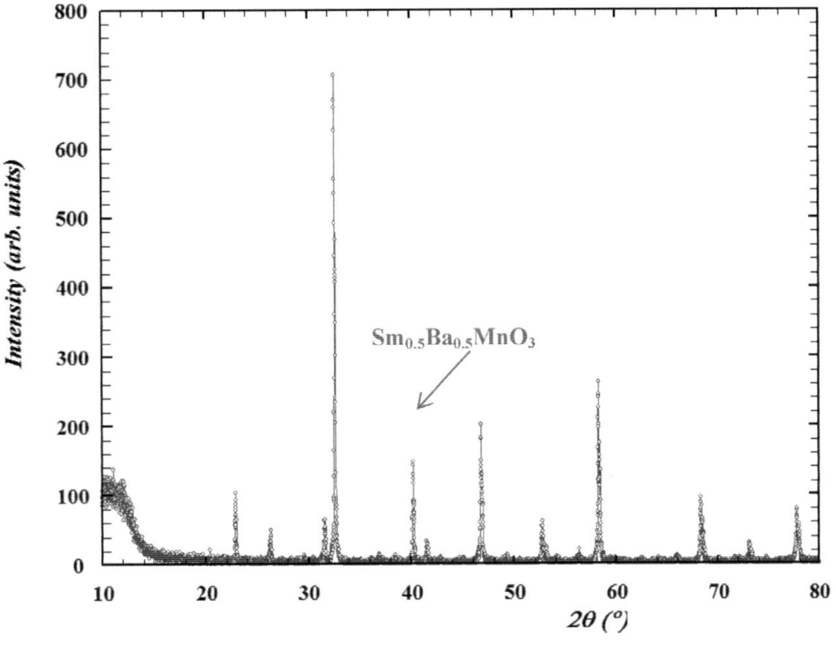

Figure 1. The X-ray diffraction (XRD) patterns show, Sm0.5Ba0.5MnO3 layered SBMO, by SIEMENS Diffraktometer D5000 Kristalloflex (Cu Kα radiation, 40 kV, 30 mA). Sm0.5Ba0.5MnO3was calcinated in air from 800 C° to 1400 C° for 25 hrs.

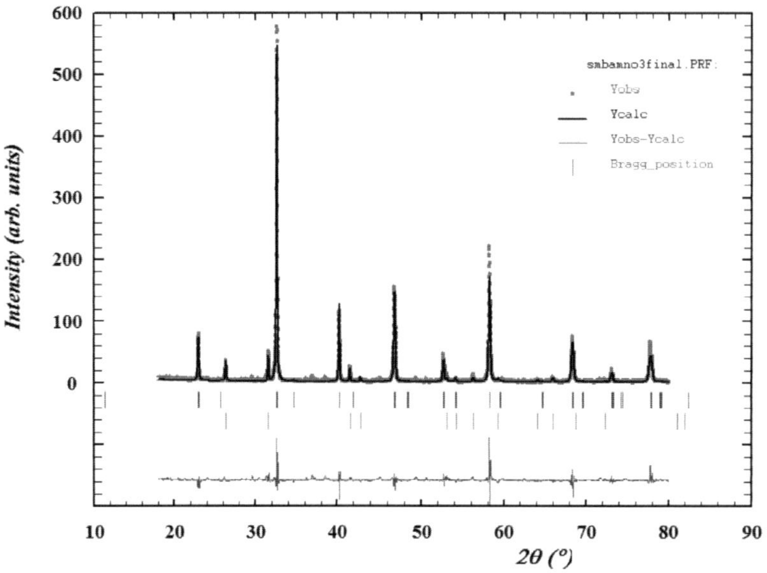

Figure.2 Rietveld refinement profile of $Sm_{0.5}Ba_{0.5}MnO_3$

Table1. Refinement results for the orthorhombic phase.

Serial No	*2Θ*	*Dhkl*	*h*	*k*	*l*
1	22.91	3.8775	0	0	1
2	32.63	2.7418	0	1	1
3	40.26	2.2387	1	1	1
4	46.82	1.9387	0	0	2
5	52.75	1.7341	0	1	2
6	58.23	1.5830	1	1	2
7	68.37	1.3709	0	2	2
8	73.08	1.2925	0	0	3
9	73.16	1.2925	2	1	2
10	7783	1.2262	1	0	3

Table 2. Refinement results from the hexagonal phase.

Serial No	*2Θ*	*Dhkl*	*h*	*k*	*l*
1	26.30	3.3848	1	0	1
2	31.576	2.8328	1	1	0

3	41.52	2.1724	0	2	1
4	42.82	2.1104	0	1	2
5	53.06	1.7239	2	1	1
6	65.88	1.4164	2	2	0

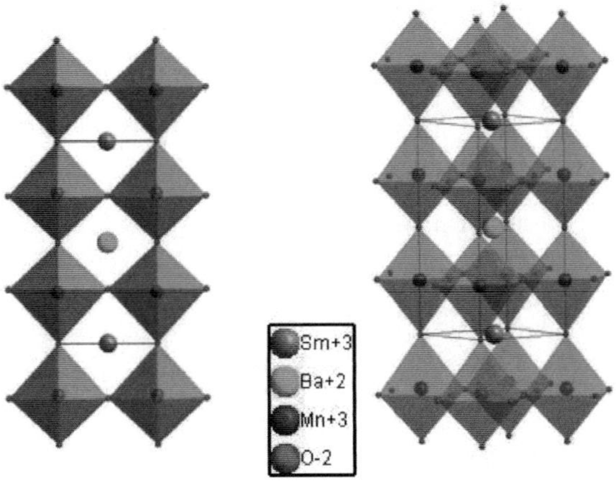

Figure.3 Schematic 3D-crystal structure of $SMBO_3$

3.2. Thermogravimetric analysis (TGA)

TGA of SBM oxide was carried out in air atmosphere with a portion of coulometric titrated sample to estimate the absolute oxygen content. Due to the heating in air, no weight loss happened from 20 - 640 °C under the ramp heating process, and the weight loss started after 640 °C, and lasted until 1000 °C as shown in Fig.4.

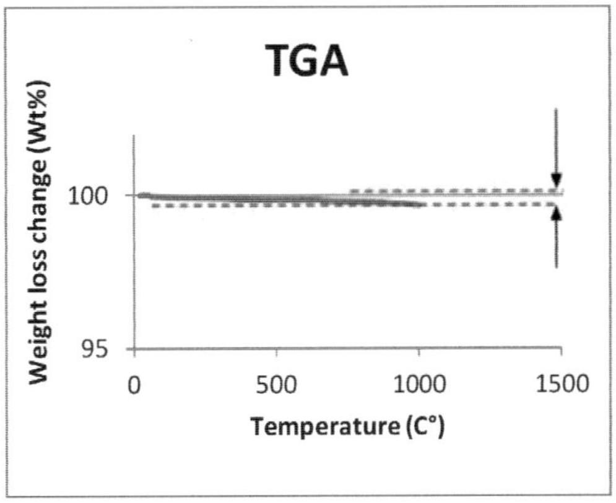

Figure.4 TGA analysis during heating process. Weight loss after 600 oC by 0.32 wt %.

A TGA analysis shows that the material is going to decrease by percentage 0.32% in heating and 0.26% in cooling stages. Apparently, the layered SBMO shows only a slight decrease or increase in weight compared to the initial weight. The analysis shows that the weight of the present material was decreased. The weight loss in the layered SBMO attributes to, the oxygen vacancy formation or decrease in oxygen content. The magnitude of weight loss by percentage after 600 °C was 0.32 wt % in heating process, and after cooling, it shows a slight weight gain for layered SBMO during cooling process and this may be due to the reoxidation of the layered SBMO sample, and it based on the delay time.

3.3. Density measurement

The density of the material was measured using both structural refinement and Archimedes theory. For structural refinement formula is:

$$\rho = \frac{Z * F_w}{V * N_a} \tag{1}$$

Where: ρ is the density, Z is the number of atoms in unit cell, F_w is the Atomic Weight [kg mol^{-1}], V is the Volume of unit cell [m^3], and N_a is the Avogadro's number [atoms mol^{-1}].

According to actual measurements from Archimedes theory, the $Sm_{0.5}Ba_{0.5}MnO_3$ shows the actual density of 6.0255 gm/cm^3. From the structural refinement the density was found to be 8.372 gm/cm^3, where the relatively percentage was about 71.98%, where it seems the particles steadily getting smaller and change in shape [21] and this suggests that SBMO oxide will show high percentages of porosity and make it as a good anode material used in SOFCs. However, high electrochemical performances tolerating carbon coking and H_2S poisons [15] depend on the B-site cations and high electrical conductivity of the pervoskite [20]. This new redox-stable mixed oxide-ion electron conductors (MIEC) anode consists of an A-site layered double and single perovskite structure, $Sm_{0.5}Ba_{0.5}MnO_{3-\delta}$ (SBMO), for hydrocarbon fuel operation. The selection of this material relied on some basic factors. Firstly, the chemical and thermal stability under fuel electrode conditions along with the A- site ordering layered SBMO [15,24]; Secondly, the oxygen kinetics and, the highest conductivity. Hence mixed-valence transition metal cations ($Mn^{+4}/Mn^{+3}/Mn^{+2}$) is supported by the layered SBMO pervoskite structure, and this results in an enhancement of the electrical conductivity and maintain a large oxygen voids content and can promote the oxygen migration in the perovskite lattices, which in terms contribute to the fast oxygen ion diffusion[21,25-28]; Third, the extremely good and active catalytic hydrogen and hydrocarbon oxidation perovskite oxides of first row transition metals containing Mn, Co and Fe-rich perovskite provide high activity in hydrocarbon oxidation[25,29,30]. Finally non-phase change and conductivity increases with regard to the addition of barium (Ba) doing with Sm and Mn [31-33].Furthermore, the existence of $BaMnO_3$ in this compound makes it a highly active and carbon tolerant anode, and also work as a good material can be used for solid oxide fuel cells.

4. Conclusion

The proper selection of SOFCs materials mainly depends on the characterization and the performance that can give in a specific application, so the synthesis of this new material makes it able to fit with the requirements needed for SOFCs. Optimum characterization and reliable analysis of material $Sm_{0.5}Ba_{0.5}MnO_3$, it might help to draw the specifications of the used SOFCs material after measuring the power density in single cell test measurements. According the characterizations performed here, it gave a good agreement with the literature results, and this indicates that this material after making a full characterization as a new A- site single pervoskite may show asuperior SOFC anode performance according to the electrochemical stability. It is expected that this single A-site and also double A- site layered SBMO anodes should give higher electricalconductivity, in addition to an excellent redox and coking tolerance and sulphur tolerance, and this will make this new series of layered SBMO oxides as a promising ceramic anode materials for SOFCs.

5. Acknowledgment

The author A.M.Abdalla is grateful to the Universiti Brunei Darussalam for providing him Graduate Research Scholarship to do PhD. We also acknowledge the help from the Centre for Advanced Materials and Energy Sciences (CAMES) for their help.

6. References

[1] San P J and Sie C 2004 A review of anode materials development in solid oxide fuel cells *J. Material .Sci* **39** 4405– 4439
[2] Nigel P B 2007 Materials Engineering for Solid Oxide Fuel Cell Technology *J.Mater. Sci. Forum* **539** 20–27
[3] Neelima M, Amitava B, Alka G, Shobit O and Kantesh B 2015 Progress in material selection for solid oxide fuel cell technology: A review" *J.Progress in Materials Science* **72** 141-337
[4] Zhu W and Deevi S 2003 A review on the status of anode materials for solid oxide fuel cells *J.Mater. Sci* **362** 228–239
[5] Z Shao,W Zhou and Z Zhu 2012 Advanced synthesis of materials for intermediate-temperature solid oxide fuel cells *J.Progress in Materials Science* **57** 804-874.
[6] Jan R, and Wolfgang G B, 2008 Trends in catalytic activity for SOFC anode materials *J. Solid State Ionics* **178** 1694–1700
[7] Norbert H M, Frank T, Sven U, Hans P B, and Detlev S 2010 Materials and manufacturing technologies for solid oxide fuel cells *J. Mater. Sci* **45** 3109–3135
[8] Ghosh A, Abul K A, and John T S I 2011 Study of Ga Doped LSCM as an Anode for SOFC *J.ECS Transac.* **35** 1337-1343
[9] Abul K A, Jung H K and John T S I 2014 Structural,electrochemical and magnetic characterization of the layered-type $PrBa_{0.5}Sr_{0.5}Co_2O_{5+d}$ perovskite *J.Solid State Chem* **213** 268–274
[10] Abul K A and John T S I, 2008 Location of Deuterium Positions in the Proton-Conducting Perovskite $BaCe_{0.4}Zr_{0.4}Sc_{0.2}O_{2.x}$ D_2O by Neutron Powder Diffraction *J.Chem. Mater* **21** 215-222
[11] Frank T, Hans P B and Detlev S 2006 10 years of Materials Research for Solid Oxide Fuel Cells At Forschungszentrum Jülich *J. Electroceramics* **17** 701–707
[12] Shabana P S S, Andanastuti M and Mahendra R S 2015 A review of anode materials development in solid oxide fuel cells *J.Renewable and Sustainable Energy Reviews* **51** 1–8

[13] Sivaprakash S, Sihyuk C, Areum J, Tae H S, Young W J, Hu Y J, Jeeyoung S, John T S I and Guntae K 2014 Layered oxygen-deficient double perovskite as an e cient and stable anode for direct hydrocarbon solid oxide fuel cells *J.Nature Materials* **4166** 1–5

[14] Abul K A, Jung H K and John T S I 2011 Structure–property relationship in layered perovskite cathode $LnBa_{0.5}Sr_{0.5}Co_2O_{5+\delta}$ (Ln= Pr, Nd) for solid oxide fuel cells *J.Power Sources*, **196** 7333-7337

[15] Abul K A and John T S Irvine 2011 Characterization of $YSr_2\ Fe_3O_{8-\delta}$ as electrode materials for SOFC *J.Solid State Ionics* **192** 225-228

[16] Hirota K, Hatta H, Yoshinaka M I M and Yamaguchi O 2003 Formation, densification, and electrical conductivity of air-sinterable $(Sm_{1-x}Ca_x)CrO_3$ *J.Mater. Sci. Eng* B **8** 3431–3435

[17] Tolstopyatova A A, Chuan Y C and Gorshkova L S 1964 Catalytic properties of samarium with respect to the dehydrogenation and dehydration of alcohols and the dehydrogenation of tetralin *J.of Russien Bulltein* **4** 7–10

[18] Hussein M A G, Buttrey D J, DeSanto P, Abd-Elgaber A A Roshdy H and Myhoub A Y Z, 2003 Formation and characterization of samarium oxide generated from different precursors *J.Acta thermomechanica* **402** 27–36

[19] Wei L, Yanyi L, Bin L, Taylor D S, Xi W and Wei P 2010 Ceria (Sm3+, Nd3+) carbonates composite electrolytes with high electrical conductivity at low temperature *J.Composits. Sci. Technol* **70** 181–185

[20] Syed M B and Javier B G 2011 Surface and redox chemistry of $Sm_{0.95}Ce_{0.05}Fe_{1-x}Ni_xO_{3-\delta}$ perovskites *J.Solid State Ionics* **194** 33–40

[21] Rada S, Pascuta P, Rada M and Culea E 2011 Effects of samarium (III) oxide content on structural investigations of the samarium–vanadate–tellurate glasses and glass ceramics *J.Non. Cryst. Solids* **357** 3405–3409

[22] Yarong W, Toshiyuk M, Ji-Guang L and Yoshiyuk Y 2011 Low-temperature fabrication and electrical property of 10 mol% Sm_2O_3 -doped CeO_2 ceramics *J.Sci. Technol. Adv. Mater* **4** 229–238

[23] Dengjie C, Ran R, Kun Z, Jun W and Zongping S 2009 Intermediate-temperature electrochemical performance of a polycrystalline $PrBaCo_2O_{5+\delta}$ cathode on samarium-doped ceria electrolyte *J.Power Sources* **188** 96–105

[24] Trukhanov S V, Trukhanov A V Szymczak H, Szymczak R and Baran M 2006 Thermal stability of A-site ordered $PrBaMn_2O_6$ manganites *J.Phys. Chem. Solids* **67** 675–681

[25] Mónica V S, Aracely M, Tulio M, Rosana Z D, Gustavo A L, Matias D A, Célia F M, Paola G M and Gilles H G 2014 Barium-modified NiO–YSZ/NiO–GDC cermet as new anode material for solid oxide fuel cells (SOFC) *J.Solid State Ionics* **261** 36–44

[26] Taskin A A, Lavrov A N, and Ando Y 2007 Fast oxygen diffusion in A-site ordered perovskites *J.Prog. Solid State Chem* **35** 481–490

[27] Khakpour Z, Yuzbashi A, Maghsodipour A and Ahmadi K 2012 Electrical conductivity of Sm-doped CeO2 electrolyte produced by two-step sintering *J. Solid State Ionics* **227** 80–85.

[28] Tanapol C, Manop P, Nattawut C and Bussarin K 2013 Fabrication of Samarium Doped Ceria Electrolyte on Rough Glass Substrate with High Electrical Conductivity by Electrostatic Spray Deposition for Intermediate Temperature Solid Oxide Fuel Cells *J. Energy Procedia* **34** 471–478

[29] Ran R, Wu X, Weng D and Fan J 2013 Oxygen storage capacity and structural properties of Ni-doped LaMnO3 perovskites *J. Alloys Compd* **577** 288–294.

[30] Nemanja D, Adrien V, Jing-Li L, Karl T Chuang, Rob H and Alan R S 2010 Correlation of Fuel Cell Anode Electrocatalytic and ex situ Catalytic Activity of Perovskites $La_{0.75}Sr_{0.25}Cr_{0.5}X_{0.5}O_{3-\delta}$ (X = Ti, Mn, Fe, Co) *J.Chem. Mater* **22** 957–965

[31] Meng L, Bin H, San P J, Jian P, Bo C and Li J2014 $BaZr_{0.1}Ce_{0.7}Y_{0.1}Yb_{0.1}O_{3-\delta}$ as highly active and carbon tolerant anode for direct hydrocarbon solid oxide fuel cells *Int. J. Hydrogen Energy* **39** 15975–15981

[32] Lay E, Dessemond L and Gauthier G 2013 Ba-substituted LSCM anodes for solid oxide fuel cells *J.Power Sources* **221** 149–156

[33] James J, Sankar M, Kumar S S and Nair K V 2004 Preparation and properties of $Ba_{2-x}Sr_xSmTaO_6$ (x=0–2): a group of new perovskite materials *J.Mater. Chem. Phys* **83** 328–333

An Effective Acid Combination for Enhanced Properties and Corrosion Control of Acidizing Sandstone Formation

Mian Umer Shafiq[1], **Hisham Khaled Ben Mahmud**[1]

[1]Petroleum Engineering Department, Curtin University, Sarawak, Malaysia

E-mail: mian.umer@curtin.edu.my

Abstract. To fulfill the demand of the world energy, more technologies to enhance the recovery of oil production are being developed. Sandstone acidizing has been introduced and it acts as one of the important means to increase oil and gas production. Sandstone acidizing operation generally uses acids, which create or enlarge the flow channels of formation around the wellbore. In sandstone matrix acidizing, acids are injected into the formation at a pressure below the formation fracturing pressure, in which the injected acids react with mineral particles that may restrict the flow of hydrocarbons. Most common combination is Hydrofluoric Acid – Hydrochloric with concentration (3% HF – 12% HCl) known as mud acid. But there are some problems associated with the use of mud acid i.e., corrosion, precipitation. In this paper several new combinations of acids were experimentally screened to identify the most effective combination. The combinations used consist of fluoboric, phosphoric, formic and hydrofluoric acids. Cores were allowed to react with these combinations and results are compared with the mud acid. The parameters, which are analyzed, are Improved Permeability Ratio, strength and mineralogy. The analysis showed that the new acid combination has the potential to be used in sandstone acidizing.

1. Introduction

The permeability of sandstone formation can be increased by applying matrix acidizing operation. To remove the damage in the vicinity of the near wellbore, sandstone acidizing has been applied for many years. The reaction mechanism between the different acids and sandstone rock is still not fully identified and understood. According to different companies 25 to 30% of the sandstone acidizing treatment fails due to lack of understanding [1]. Mud acid (HF:HCl) has been developed and applied on sandstone formation since 1965 [2, 3]. Lake in 1962 [4] studied the effect by varying the concentration of HCl in mud acid to study effects on the carbonate particles. Lake et al. also studied the effect of different HF concentrations during treatments. Different acid combinations have been tried on sandstone formation to solve the problems caused by mud acid such as less penetration, fast spending and precipitation etc. HF acid is the essential constituent of all developed acid because of the presence of fluoride ion (F^-). This is the only ion discovered so far which can reacts with quartz which resulted in the removal of damage around the wellbore. "Therefore, the goal of acidizing in sandstone formation is to dissolve minerals that present in pores to enhance the formation permeability [5]".

2. Mechanism of Sandstone Acidizing

Acid soluble minerals present around the wellbore start dissolving in HF acid as soon as acid enters the formation. The reactivity of fluoride ion makes this acid very reactive, that's why the reaction is very fast and most of the acid spends at early stages [6, 7]. To slow down this fast reaction rate, HF acid is mixed with another mineral acid (HCl or HCOOH), which acts as a buffer acid and this buffer effect allows deeper penetration of acid into the formation [8]. Sandstone formation does not react with acids like nitric and sulphuric effectively [2]. During acidizing with mud acid, the most important concerns are the reaction of carbonates with HF and HCl and HF with silica. The reactions that take place when HF acid enters the formation are [9, 10]:

$$4HF + SiO_2 \longrightarrow SiF_4 + 2H_2O \quad (1)$$
$$SiO_2 + 6HF \longrightarrow H_2SiF_6 + 2H_2O \quad (2)$$
$$H_2SiF_6 + 4H_2O \longrightarrow Si(OH)_4 + 6HF \quad (3)$$

Other precipitates formed due to this reaction are:

$$H_2SiF_6 + Na+ \longrightarrow Na_2SiF_6 + H_2 \quad (4)$$
$$H_2SiF_6 + K+ \longrightarrow K_2SiF_6 + H_2 \quad (5)$$
$$2HF + Ca+ \longrightarrow CaF_2 + H_2 \quad (6)$$

3. Development of New Acids

To avoid the precipitation of above products, new acid systems and acidizing techniques were developed by different researchers. To avoid the precipitation of K_2SiF_6, Na_2SiF_6, and CaF_2; HCl acid is injected ahead of HF acid to remove the ions of calcium, potassium and sodium [11]. To avoid the silica precipitation $Si(OH)_4$, hydrochloric or formic acid have been added with HF acid. This addition has allowed slower reaction of acid with the formation and also this acid maintains the pH of the reaction products, thus avoiding silica precipitation to some extent [12]. Still, no research has been done to change HF acid with some other acids consists of fluoride ion. This paper presents new combinations of acids for main acid stage of sandstone acidizing. One combination consists of HF acid and phosphoric acid in which phosphoric acid is the replacement of HCl acid in mud acid mixture. Second combination consists of fluoboric and formic acid in which fluoboric acid is replacing hydrofluoric acid in organic mud acid mixture. Different tests are performed using this acid and analysis is presented.

4. Experimental Work

Acid saturation was performed on Berea core samples of 3 inch in length and 1.5 inch in diameter at standard temperature and pressure condition. Properties of sandstone core sample are presented below:

- Company: Cleveland Querries
- Weight: 0.9 kg/core
- Effective Porosity: 9 – 10%
- Permeability: 65 – 70%
- Strength: 53.1 MPa
- Mineralogy: 93% quartz, 4.3% clay, 1.75% carbonates, 0.95% iron compounds
- 1.5" Diameter, 3" Length

4.1. Acid Volume Calculation

Volume of acid is calculated using the formula:

$$M_1V_1 = M_2V_2 \quad (7)$$

Concentrations of acids are already given by the supplier (acids in liquid form), mentioned in table 1.

IOP Conf. Series: Materials Science and Engineering **121** (2016) 012002 doi:10.1088/1757-899X/121/1/012002

Table 1. Acid Combinations

Sr. No	Acids	Concentration
A	HCl	37%
B	H_3PO_4	100%
C	HF	48%
D	CH_3COOH	100%
E	HCOOH	100%
F	HBF_4	50%

For example, the calculation for 3%HF: 12% HCl is as follows (Table 2).

Table 2. Acid Volume Calculation

HF	HCl
$M_1V_1 = M_2V_2$	$M_1V_1 = M_2V_2$
(48) V1 = (3) (175)	(37) V1 = (12) (175)
V1 = 10.75 ml	V1 = 56.75 ml
The total amount of solution is 175ml; the remaining volume will be filled in by the distilled water. Volume of water = 175-56.75-10.75= 107.5 ml	

4.2. Experimental Setup

Different experiments were performed on the core samples before and after acidizing process. These experiments consist of permeability, strength and FESEM analysis. Figure 1 shows the experimental setup for saturation of core sample.

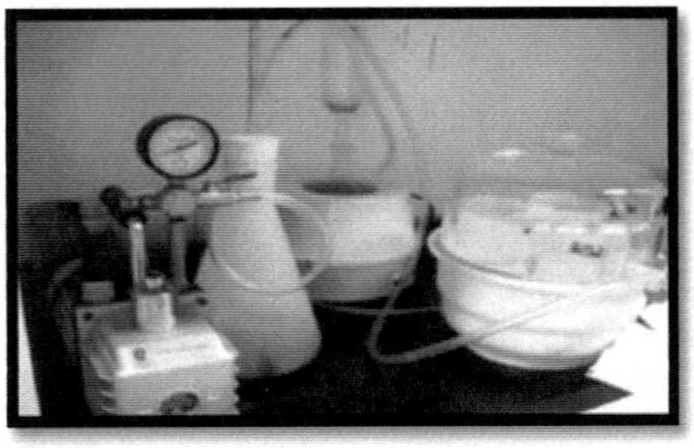

Figure 1: Core saturation experimental setup

4.3. Saturation of Core Sample

- Prepare the desiccators and vacuum pump.
- Immerse the sample in 175 ml of acid solution.
- Place the lid on the desiccators and open the tap.
- Place this saturation system in appropriate size fume cupboard.
- Connect the tap to the vacuum pump and open the tap slowly to evacuate the desiccators.
- When vacuum is achieved in desiccators, close the tap, disconnect the vacuum supply and wait 4 hours to allow the core to react with the acid.

IOP Conf. Series: Materials Science and Engineering **121** (2016) 012002 doi:10.1088/1757-899X/121/1/012002

- After saturation, keep the sample in the oven for 24 hours to evaporate the acid present in the sample.

Tests are performed on core sample before and after acidizing process. Figure 2 represents the flow chart of all the experiments performed.

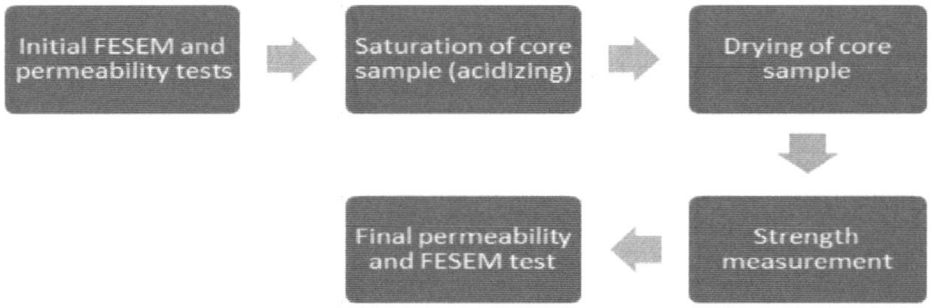

Figure 2: Flow Diagram of Experiments

5. Results

During acidizing, core samples were allowed to react with different acid combinations. After the reaction, different properties of these core samples were measured and analyzed as follows:

5.1. Improved Permeability Ratio

The Improved Permeability Ratio (IMPR) is the ratio of the final value of permeability to its initial value. From table 3 and figure 3, it is clear that increase in the permeability when 3%HF: 9% H_3PO_4 is used gives much better results than mud acid. Increase in permeability is almost 2.5 folds when 3%HF: 9% H_3PO_4 is applied to sandstone formation while with mud acid the increase is 2 folds. But when HBF_4: HCOOH is used all the ratios of acids used show better results than mud acid, which indicates this acid combination is very effective in sandstone acidizing. Therefore, it can be concluded from permeability results that these combinations are better than mud acid.

Table 3. IMPR Values

Sr. No	Acids	Initial K	Final K	IMPR
A	3%HF: 12%HCl	70.26	141.76	2.01
B	3%HF: 12%H_3PO_4	71.46	131.47	1.83
C	3%HF: 9% H_3PO_4	70.5	165.90	2.35
D	1.5%HF: 9% H_3PO_4	72.89	115.29	1.58
E	3%HBF_4: 12%HCOOH	71.81	164.96	2.29
F	3%HBF_4: 9%HCOOH	71.82	162.32	2.26
G	1.5%HBF_4: 9%HCOOH	71.78	140.25	1.95

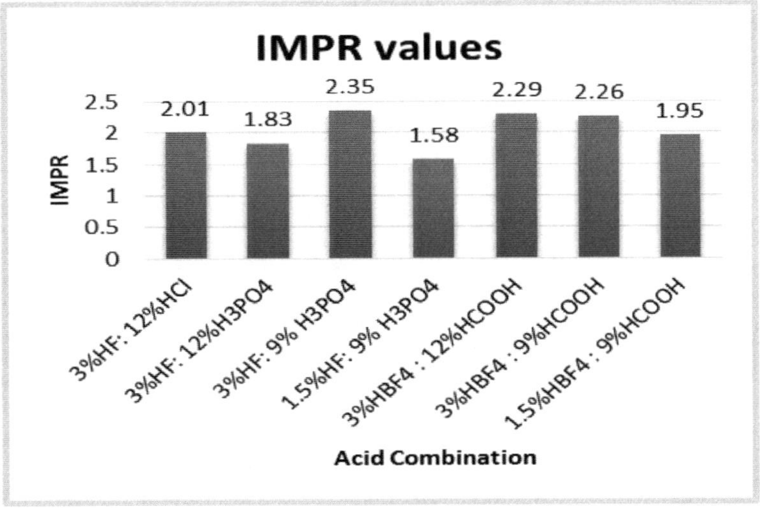

Figure 3. IMPR Values.

5.2. Compressive Strength Analysis

To find the compressive strength of the core samples all reacted samples are utilized. The initial value is kept constant as it can be measured only once with an unreacted core sample. The combination (HF: H_3PO_4), which shows maximum increase in the permeability; it causes change in strength of the core sample about 25% which is almost equal to the result of mud acid. The results are presented in table 4 and figure 4. The strength change with mud acid is also 25% but the change in permeability is less as compare to other acids used. This is due to very fast reaction of the HF acid with the core. Precipitates may forms due to this fast spending of acid which ultimately blocks the pore spaces. This blockage is one of the main cause of less increase in permeability as desired. 3%HBF_4: 12%HCOOH causes the maximum change in strength which is 36%. This combination also leads to more increase in permeability, which indicates that this acid although changes permeability, it is also deteriorating the rock by dissolving the cementing minerals.

Table 4. Compressive Strength Values

Sr No	Acid Combination	Initial Peal Load (Mpa)	Final Peak Load (MPa)	% change
A	3%HF: 12%HCl	53.1	39.6	25.42
B	3%HF: 12%H_3PO_4	53.1	42.7	19.59
C	3%HF: 9% H_3PO_4	53.1	39.5	25.61
D	1.5%HF: 9% H_3PO_4	53.1	41.7	21.47
E	3%HBF_4: 12%HCOOH	53.1	33.7	36.54
F	3%HBF_4: 9%HCOOH	53.1	38.4	27.68
G	1.5%HBF_4: 9%HCOOH	53.1	47.7	10.17

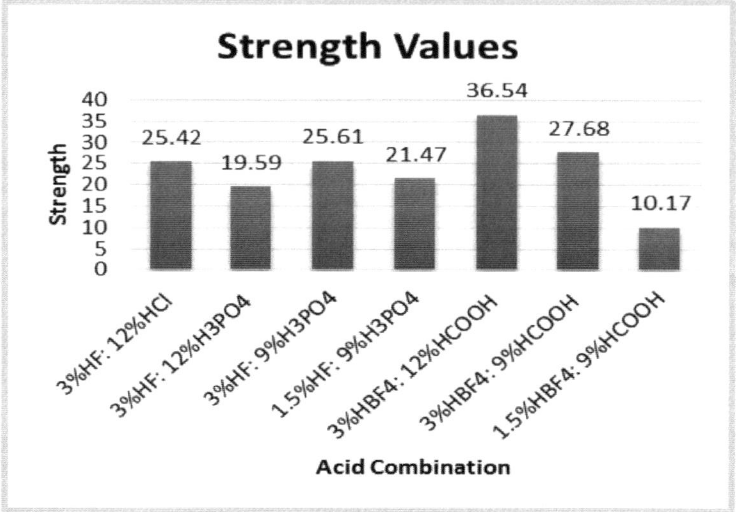

Figure 4. Compressive Strength

5.3. Mineralogy Estimation

The Field Emission Scanning Electron Microscopy (FESEM) is a versatile, non-destructive technique that reveals detailed information about the morphology and presence of different minerals present in the sample before and after the acidizing operation. The mineralogy changes will indicate that which combination is dissolving more minerals. Mainly two tests are performed using FESEM (EDX and Morphology). Details of EDX are illustrated in table 5. Mineralogy analysis can help in the determination of corrosive nature of the acid which is one of the main theme of this research. From reacted core samples, the cores which were reacted with best ratio of acid combinations were selected for mineralogy test i.e. 3%HF: 12%HCl, 3%HF: 9%H_3PO_4, 3%HBF_4: 12%HCOOH.

Table 5. Mineralogy Values

	Initial composition		After Main Acid Reaction		
	Elements	% weight	HF + HCl	HF + H_3PO_4	HBF_4 + HCOOH
1	Oxygen (O)	55.90	51.03	50.97	55.96
2	Silicon (Si)	37.80	27.02	34.27	36.35
3	Aluminium (Al)	2.60	1.89	2.49	2.30
4	Potassium (K)	1.00	0.90	0.86	0.95
5	Iron (Fe)	0.95	1.28	0.78	2.36
6	Fluorine (F)	0.00	0.00	7.28	0.00
7	Phosphorus (P)	0.00	0.00	1.12	0.00
8	Chlorine (Cl)	0.00	1.17	0.87	0.96
9	Calcium (Ca)	1.75	0.81	1.36	1.12
10	Carbon (C)	0.00	15.92	0.00	0.00

6. Discussion and Analysis

The results based on conducting three tests (permeability, mineralogy, strength) stated that the two combinations 3%HF: 9%H_3PO_4 and 3% HBF_4: 12%HCOOH showed the best results with consistency. The combination 3%HBF_4: 9%HCOOH also show better results and values are near to mud acid results. Therefore, the two final best combinations are 3%HF: 9%H_3PO_4 and 3%HBF_4: 12%HCOOH.

When HF: HCl reacted with the core sample, there is a change in the mineralogy where carbon is formed as seen from table 5. The change in permeability is also less when this combination (HF: HCl) is used. Carbon is not formed when core samples reacted with other combinations (HF: H_3PO_4, HBF_4: HCOOH) used in this study. This also indicates that the presence of carbon component may be a reason of less permeability increment when mud acid was used.

Increase in iron components was observed when (HF:HCl and HBF_4:HCOOH) are used for acidizing. This indicates that acids are corrosive and corrosion inhibitor during acidizing may be needed when these acids are injected into the well. But when (HF: H_3PO_4) was applied to sandstone formation, the iron component decreased which, is opposite to other combinations. Thus, corrosion inhibitor in not needed when this new acid combination is used which reduces the cost of operation.

The results of strength tests are also in accordance with the results of permeability. It is observed that the acid which cause more change in permeability also causes more decrease in the compressive strength and vice versa. The maximum amount of silicon has been dissolved when hydrochloric acid was used (HF: HCl) mentioned in Table 5. The minimum amount of silicon has been dissolved when formic acid was used (HBF_4: HCOOH) and when phosphoric acid has been used (HF: H_3PO4), the results are in between the combinations in which HCl and HCOOH are used. The combinations in which HCOOH is present cause more change in strength of the core compared to other combinations. Mineralogy results reveals that the combinations (HF: H_3PO_4, HBF_4: HCOOH) dissolves less amount of silicon minerals and causes more change in strength of the sample. Therefore, it can be concluded that silicon is not the cementing material of the core samples used in this research.

By considering all the results and analysis it can be demonstrated that HF: H_3PO_4 is the best combination from all the combinations applied on sandstone formation in this paper. It causes more increase in permeability, less change in strength of the core sample and causes less corrosion of the materials. More increase in permeability is also the indication of more penetration of the acid into the formation which is favorable for acidizing process.

7. Conclusion

The change in permeability due to acidizing is better when hydrochloric acid is replaced by phosphoric acid in main acid stage. Phosphoric acid acts as a buffer acid and reduces the corrosion rate of the pipeline as analyzed from the mineralogy test. This buffer effect is also responsible for more increase in permeability and deep penetration. It can be concluded from this analysis that corrosion inhibitor may not be required when phosphoric acid is used as it acts as a corrosion inhibitor itself.

References

[1] Bybee K 2001 A Structured Approach to Candidate Selection and Stimulation Treatment Design *Society of Petroleum Engineers* SPE-0801-0059-JPT

[2] Smith C F and A R Hendrickson 1965 Hydrofluoric Acid Stimulation of Sandstone Reservoirs *Society of Petroleum Engineers* SPE-980-PA

[3] Abdelmoneim S S and H A Nasr-El-Din 2015 Determining the Optimum HF Concentration for Stimulation of High Temperature Sandstone Formations *Society of Petroleum Engineers* SPE-174203-MS

[4] Walsh M P, L W Lake and R S Schechter 1982 A Description of Chemical Precipitation Mechanisms and Their Role in Formation Damage During Stimulation by Hydrofluoric Acid *Society of Petroleum Engineers* SPE-10625-PA
[5] Schechter R S 1992 *Oil Well Stimulation* Englewood Cliffs NJ: Prentice Hall
[6] Gdanski R 1998 Kinetics of Tertiary Reactions of Hydrofluoric Acid on Aluminosilicates *Society of Petroleum Engineers* SPE-31076-PA
[7] Al-Dahlan M N, H A Nasr-El-Din and A A Al-Qahtani 2001 Evaluation of Retarded HF Acid Systems *Society of Petroleum Engineers* SPE-65032-MS
[8] Yang F, H A Nasr-El-Din and B M Al-Harbi 2012 Acidizing Sandstone Reservoirs Using HF and Formic Acids *Society of Petroleum Engineers* SPE-150899-MS
[9] Shaughnessy C M and K R Kunze 1981 Understanding Sandstone Acidizing Leads to Improved Field Practices *Society of Petroleum Engineers* SPE-9388-PA
[10] Al-Harbi B G, M H Al-Khaldi and K A AlDossary 2011 Interactions of Organic-HF Systems with Aluminosilicates: Lab Testing and Field Recommendations *Society of Petroleum Engineers* SPE-144100-MS
[11] Hill A D, K Sepehrnoori and P Y Wu 1994 Design of the HCl Preflush in Sandstone Acidizing *Society of Petroleum Engineers* SPE-21720-PA
[12] Yang F, H A Nasr-El-Din and B A Harbi 2012 Acidizing Sandstone Reservoirs Using HF and Organic Acids *Society of Petroleum Engineers* SPE-157250-MS

IOP Conf. Series: Materials Science and Engineering **121** (2016) 012003 doi:10.1088/1757-899X/121/1/012003

Software Defined Networking challenges and future direction: A case study of implementing SDN features on OpenStack private cloud

Veeramani Shamugam[1*], I Murray[2], J A Leong[3] and Amandeep S Sidhu[3]
[1]Department of Electrical and Computer Engineering, Faculty of Engineering and Sciences, Curtin University Sarawak Campus, Miri, Malaysia.
[2]Department of Electrical and Computer Engineering, Faculty of Science and Engineering, Curtin University, Bentley, Perth, Western Australia, 6102.
[3]Curtin Sarawak Research Institute, Curtin University Sarawak Campus, Miri, Malaysia.
*Corresponding author: s.veeramani@curtin.edu.my

Abstract— Cloud computing provides services on demand instantly, such as access to network infrastructure consisting of computing hardware, operating systems, network storage, database and applications. Network usage and demands are growing at a very ***fast*** rate and to meet the current requirements, there is a need for automatic infrastructure scaling. Traditional networks are difficult to automate because of the distributed nature of their decision making process for switching or routing which are collocated on the same device. Managing complex environments using traditional networks is time-consuming and expensive, especially in the case of generating virtual machines, migration and network configuration. To mitigate the challenges, network operations require efficient, flexible, agile and scalable software defined networks (SDN). This paper discuss various issues in SDN and suggests how to mitigate the network management related issues. A private cloud prototype test bed was setup to implement the SDN on the OpenStack platform to test and evaluate the various network performances provided by the various configurations.

1. Introduction

A cloud is defined as a place, over network infrastructure, where information technology (IT) and computing resources such as computer hardware, operating systems, networks, storage, databases and even entire applications are available instantly, on demand [2-3]. The new cloud architecture provides network infrastructure as a service through a group of layers, which are also provided as services. The cloud framework consists of three main layers: 1. Software as a Service (SaaS) - allows multiple end users to access applications which are running on cloud infrastructure through a web browser. The end users do not manage or control the software in the cloud. 2. Platform-as-a-Service (PaaS) - allows the users to install applications developed through programming languages or packages on the cloud but they do not control or manage the cloud. 3. Infrastructure as a service (IaaS) - provides users with processors, storage, networks and other computing resources as a service and the user do not have to control or manage the infrastructure but they do have control over the operating systems (OS), applications and programming frameworks [1].

2. Software Defined Networking (SDN)

2.1 Background

Software defined networking, referred to as a revolutionary new idea in computer networking, is the new approach that promises to dramatically simplify network-control and the management plane. It is achieved through innovative network programmability. In the cloud environment, monitoring network users and the resources they access, managing the infrastructure, as well as supporting demand becomes complicated [4-5]. Networking devices have control and data movement functions on the same device.

Traditionally, the network admin has to control from the network management plane in order to configure each network's devices separately. This static setup of current network devices does not permit control-plane configuration. The Software-defined networking (SDN) approach allows for open, user-controlled, management of the forwarding hardware in network elements [6]. SDN implements centralized control-plane intelligence while keeping the data plane separate, which in turn enables the administrator to configure network hardware directly from the controller. This approach of centralizing control of the entire network makes the network highly flexible [7-8]. To meet growing demand and cater for network instability, Manar Jammal proposed Software Defined Networking (SDN) combined with network function virtualization (NFV) [6]. SDN isolates the network control logic from the hardware which in turn enables the network administrator to have more control over network functions and a global view of networks.

2.2 Software Defined Networking Architecture

Traditional networks are difficult to automate because of the distributed nature of their decision making process for switching or routing which are collocated on the same device. Managing complex environments using traditional networks is time-consuming and expensive, especially in the case of generating virtual machines, migration and network configuration. The demand for service and network usage is growing rapidly. The current nature of data such as video traffic, big data centers and mobility of network users pose significant challenges to network operators who are facing spectrum congestion. Also, data-center operators are facing tremendous growth in the number of servers and virtual machines and increasing server-to-server communication traffic. To overcome these challenges, network operators require efficient, flexible, agile and scalable software defined networks (SDN). The key role of a SDN is to aggregate and centralize the control plane which is a promising solution for network management and control problems as shown in figure 1.

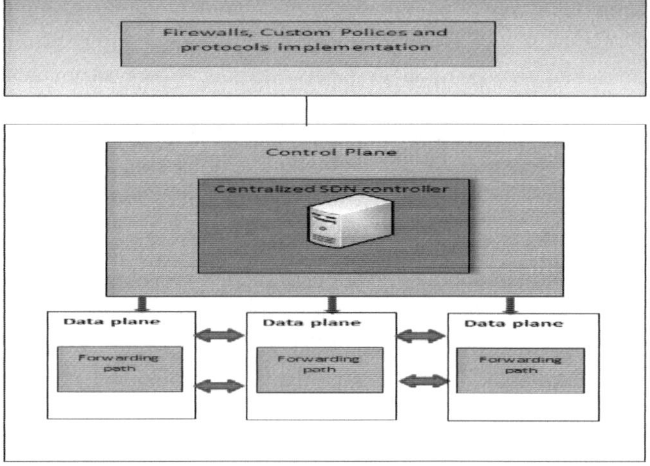

Figure 1: SDN architecture.

An SDN consists of three sections:

1. The Network Management Centre is responsible for implementing various functions such as firewalls, custom policies and protocol implementations.
2. The control plane function centralizes the control plane intelligences (switching and routing) to the controller. This allows administrators to configure the network hardware directly from the controller. This approach makes the network highly flexible.
3. The Data plane represents packet forwarding hardware in the SDN architecture

3. Issues in SDN

Although SDN is a favorable solution for IT and cloud providers and enterprises, SDN faces some challenges that hinders its performance and implementation. The list of SDN challenges consists of: Controller placement, Scalability, Performance, Security, Interoperability and Reliability [6]. SDN controllers must be wisely configured and the SDN's network topology authenticated to prevent manual errors and increase network availability. In a traditional network when one network or many network devices fail, network data flow is routed through another or nearby nodes or devices to continue data flow continuity.

However in the centralized controller architecture of SDN, a single controller is in charge of all the networks, and if there is a failure in the central controller, the whole network collapses since there is no alternate controller. To address this issue the cloud organization needs to focus on how to efficiently utilize main controller functions that can increase network reliability. The SDN controller should have the ability to support multiple-path solutions or fast traffic rerouting to active links if there is a path/link failure. If the main controller fails, the newer architectures support an alternate controller which can handle traffic flow. Controllers also support technologies such as Virtual Router Redundancy Protocol (VRRP) and Multi-Chassis link aggregation groups to increase network availability.

The second SDN challenge is the scalability because in this approach, the data and control planes are decoupled but they can progress independently as long as the API connects them. This centralized view of the network accelerates changes in the control plane. The decoupling process has its own drawbacks, such as having a standard API for both planes and the SDN controller become the bottle neck in a situation where the network scales the number of switches and number of nodes up.

Performance of the network is another important area to look into. SDN is a flow based technique, so the performance is measured on two metrics: flow-setup time and how many flows per second the controller can switch. Flow-setup works in two modes, which are proactive and reactive. These two modes have their respective flow initiation and flow limitation overheads. To overcome the performance limitation, focus is needed on factors affecting flow-setup time and I/O performance of the controller. There are means and ways to increase the performance by considering well-known optimization techniques, such as input/output batching and using the Maestro approach which uses techniques such as input batching threshold (IBT) and pending raw packet threshold (PRT).

3.1 Network Management in SDN

Software Defined Networking and virtualization are a combined solution to overcome the challenges that is faced in traditional, legacy, networks. SDN operates on an accumulated and centralized control plane that provides a likely solution for network management and control problems. The main idea here is that SDN separates the data plane and control plane in addition to providing the flexibility of programmability to the centralized control plane. The ultimate goal of SDN is to provide open, user-controlled, management of the forwarding hardware and network elements. It functions on the idea of a centralized control-plane intelligence by keeping the data plane separate. Thus, the network hardware devices keep their switching fabric (data plane), but hand over their intelligence (switching and routing functionalities) to the controller [6]. This enables the administrator to configure the network hardware directly from the controller. This centralized control of the entire network makes the network highly flexible.

3.2 Prototype test bed to implement SDN features.

In this research project, we implemented the various features of a SDN such as a controller node responsible for collecting routing information and making routing decisions centrally. The SDN data plane will be the hypervisor (compute node) in the prototype model and the functionality of Software Defined Networking in a private cloud built using OpenStack.

4. OpenStack Frameworks

4.1 OpenStack

OpenStack is an open source cloud computing platform that can support all types of cloud environments. OpenStack offers the same open source cloud solution as you would find on Eucalyptus, and also manages to outdo Eucalyptus when it comes to support and troubleshooting. OpenStack also offers PaaS, IaaS and even NaaS in its newer releases. Besides OpenStack offering a cloud computing software solution, it also contains many additional features and a number of tools [10]. These include: Scaling in size depending on demand and user needs; and processing big data and heavy workloads with tools like Hadoop High-performance Computing (HPC) environments. OpenStack has deployed its platform with PaaS and IaaS concepts in mind and supports a wide variety of hardware including the ARM processor architecture. It transcends both services and manages to tie in neatly with the overall package offerings from other clouds, such as Windows Azure [5]. Because of its universal compatibility with most high level programming languages, it can be considered as an IaaS service model. The cloud frame work shown in figure 2 was used to execute various applications and to test our research objective.

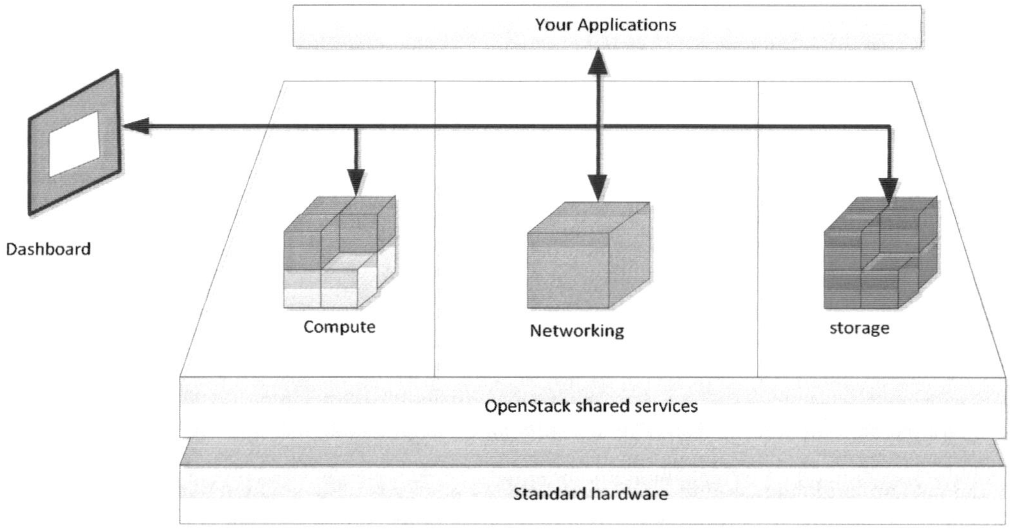

Figure 2: Open source cloud computing framework.

The open source cloud computing frame work consists of three important parts as discussed below:

Compute Node: Offers on-demand computing resources by provisioning and managing large networks of virtual machines. Compute resources are accessible via APIs for developers building cloud applications and via web interfaces for end users

Storage Node: Offers object storage and block storage, with many deployment options depending on use case.

Network Node: Offers pluggable, scalable and an API-driven system for managing networks and IP addresses. It can also be used to increase the value of existing datacenter assets. Lastly, it also ensures the network will not bottleneck a cloud deployment and gives end users real self-service, even over their network configurations [10]

4.2 Network Architecture

In this research, a basic three node architecture was implemented as shown in figure 3. This architecture provides high end computing, networking and storage facilities.

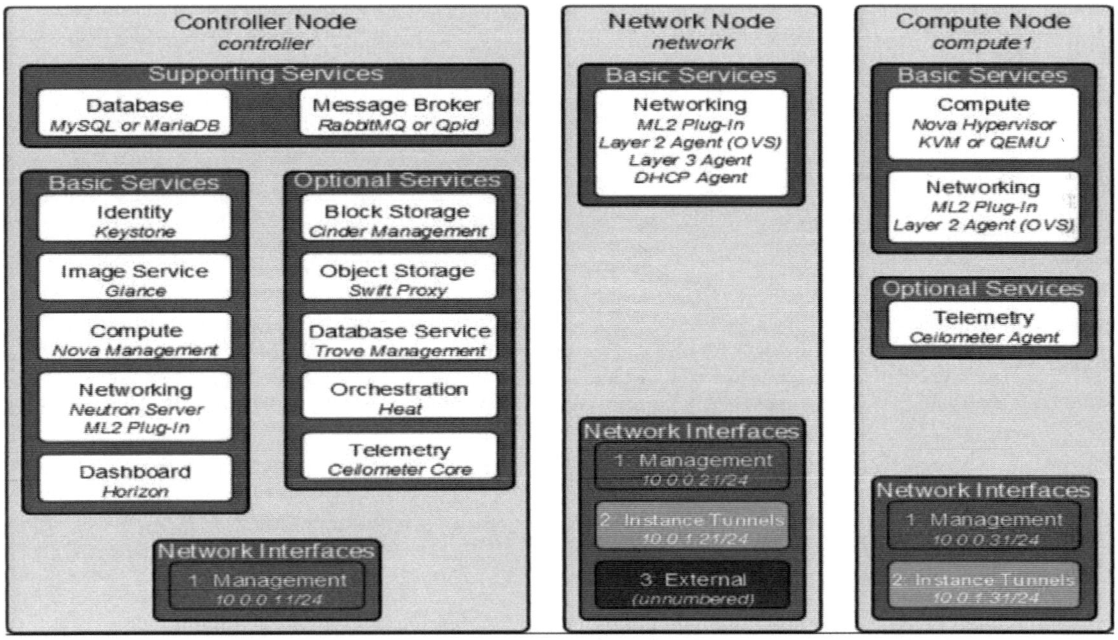

Figure 3: Architecture of three nodes in an open source cloud.

In figure 3 the first block section is the Controller node, the second is the Network node, and the third is for a Compute node.

The controller node is responsible for the following basic services: identity, image, management portion of Compute and Networking, Networking plug-in and the dashboard. It also runs additional supporting features such as a message broker, database in MySQL and Network time Protocol (NTP) [10].

The Network node executes the following services: networking ML2 plugin, layer 2 and layer 3 agents that provide and operate tenant networks. The main role of layer 3 is routing, network address translation and DHCP services. Providing virtual networks and tunnels is taken care of by layer 2.

Compute nodes execute the hypervisor part of compute services that functions as tenant virtual machines or instances. It also runs networking plug-ins and other optional services.

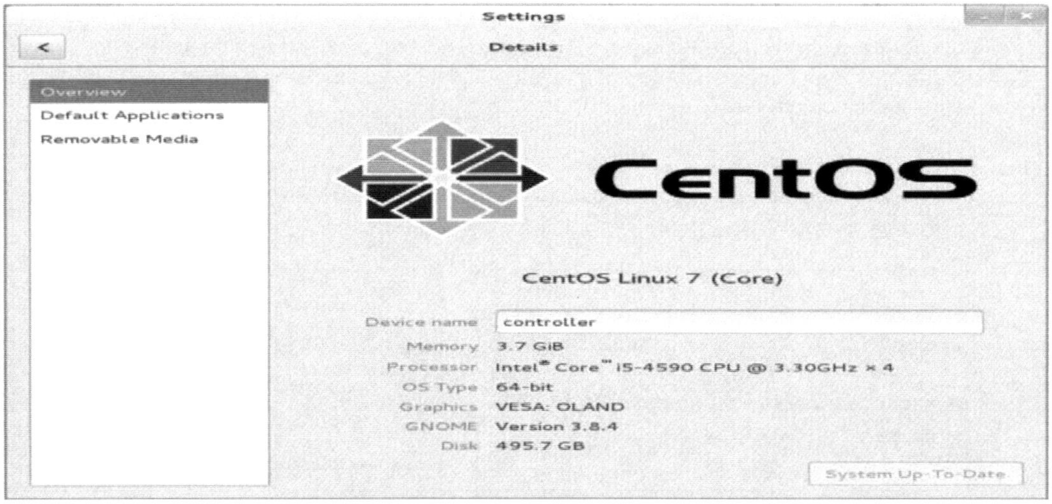

Figure 4: Installation of CentOS.

Figure 4 shows the installation CentOS. OpenStack is subsequently installed on top of CentOS, where it runs a set of software tools, building and managing cloud computing platforms for public and private clouds. The aim is to enable organizations, regardless of size, to create and offer cloud computing services running on standardized hardware. The OpenStack community has collaboratively identified nine key

Components that are part of the "core" of OpenStack, which are distributed as a part of any OpenStack system and maintained by the OpenStack community. By provisioning and managing large networks of virtual machines, these components enable enterprises and service providers to offer on-demand computing resources

4.3 Basic services: Identity service

Keystone provides an authentication and authorization service for other OpenStack services. Keystone integrates with LDAP to provide a central list of all users of the OpenStack cloud and allows administrators to set policies that control which resources various users have access to. It provides multiple means of access, meaning developers can easily map their existing user access methods against Keystone.

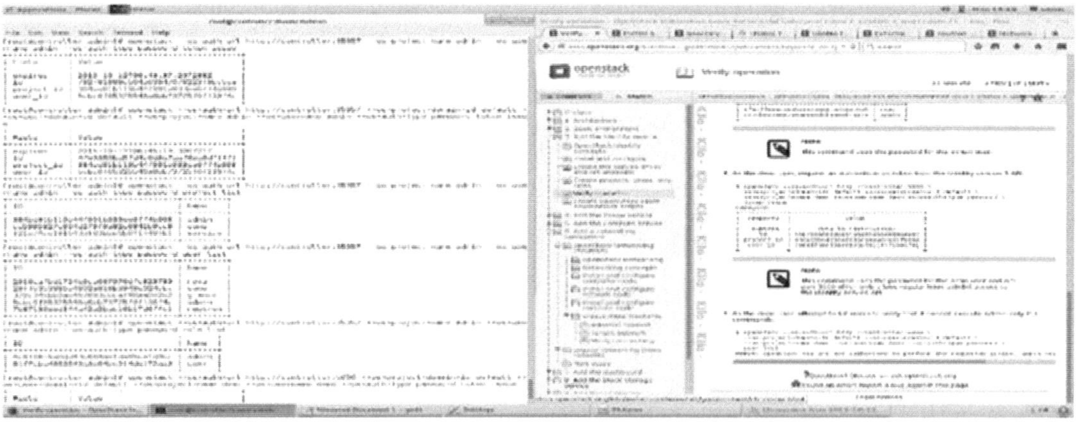

Figure 5: Implementation of the Dashboard service.

4.4 Basic services: Dashboard service

Horizon provides administrators and users a graphical interface to access, provision, and automate cloud-based resources. Developers can access all of the components of OpenStack individually through an application programming interface (API), but the dashboard provides system administrators a look at what is going on in the cloud, and to manage it as needed. It's the primary way for accessing resources if API calls are not used.

4.5 Basic services: Image service

Glance stores and retrieves virtual machine disk images. It allows these images to be used as templates when deploying new virtual machine instances. One of the main benefits to a cloud platform is the ability to spin up virtual machines when users request them. By creating templates for virtual machines, Glance helps to achieve this benefit. Also, it can copy or snapshot a virtual machine image and later on allow it to be recreated. Glance can also be used to back up existing images to save them. Glance integrates with Cinder to store the images. OpenStack Compute makes use of these stored images during instance provisioning.

4.6 Basic services: Networking

Neutron (formerly Quantum) provides the networking capability for OpenStack. It helps to ensure that each of the components of an OpenStack deployment can communicate with one another quickly and efficiently. Neutron manages the networking associated with OpenStack clouds. It is an API-driven system that allows administrators or users to customize network settings. It supports the Open Flow software defined networking protocol and plugins are available for services such as intrusion detection, load balancing and firewalls.

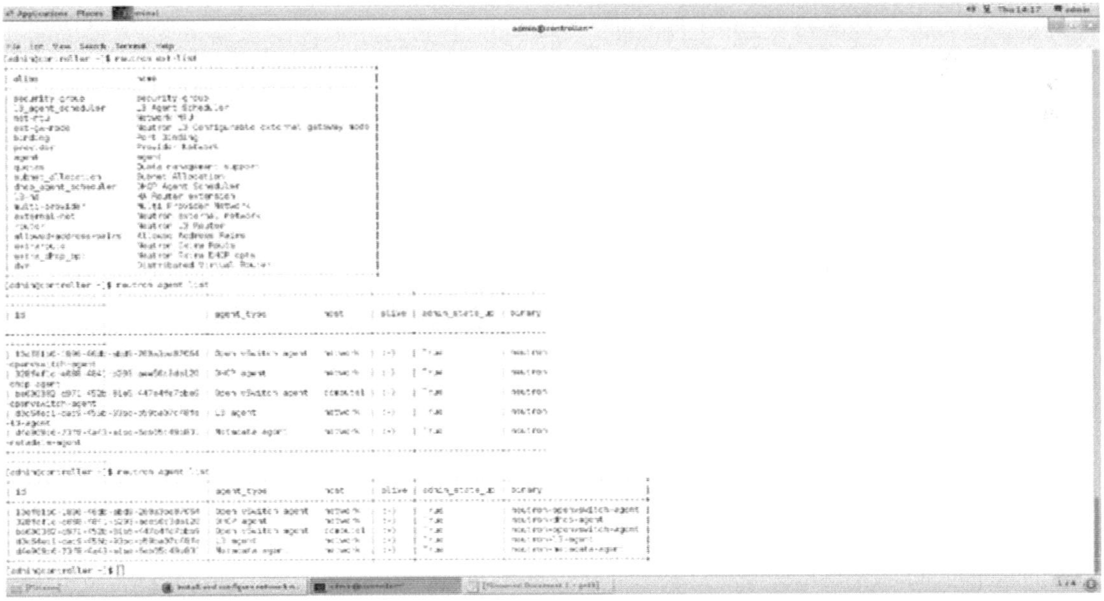

Figure 6: Implementation of the Networking service (Neutron).

4.7 Basic services: Compute service

Nova is designed to manage and automate the provisioning of compute resources. This is the core of the virtual machine management software, but it is not a hypervisor. Instead, Nova supports virtualization technologies including KVM, Xen, ESX and Hyper-V, and it can run on bare-metal and high performance computing configurations too. Compute resources are available via APIs for developers and through web interfaces for administrators and users. The compute architecture is designed to scale horizontally on standard hardware.

4.8 Basic services: Database service

Trove is the database as a service open source project for OpenStack. Trove is to provide scalable and reliable cloud database as a service, provisioning functionality for relational and non-relational database engines, and to improve its full-featured and extensible open source framework. Trove is designed to run entirely on OpenStack. Cloud users and database administrators can provision and manage multiple database instances as needed. Initially, the service will focus on providing resource isolation at high performance while automating complex administrative tasks including deployment, configuration, patching, backups, restores and monitoring.

5. Conclusion

In this paper, we present an implementation of the various SDN features such as a controller node responsible for collecting routing information and making routing decisions centrally on OpenStack. The SDN data plane will be the hypervisor (compute node) in the prototype model and functionality of the Software Defined Network is deployed in a private cloud on OpenStack.

6. References

[1] Furht B and Escalante A 2010 Handbook of cloud computing (New York: Springer)
[2] Wan Z 2010 *Cloud Computing infrastructure for latency sensitive applications* (IEEE) In Communication Technology (ICCT), 2010 12th IEEE Int. Conf. p 1399-1402
[3] Gathering Clouds of XaaS! http://ibm.com/developer
[4] Costa P, Migliavacca M, Pietzuch P and Wolf A L 2012 *NaaS: Network-as-a-Service in the Cloud* Presented as part of the 2nd USENIX Workshop on Hot Topics in Management of Internet, Cloud, and Enterprise Networks and Services 2012.
[5] Dudkowski D, Tauhid B, Nunzi G and Brunner M 2011 *A prototype for in-network management in NaaS-enabled networks* (IEEE) In Integrated Network Management (IM), 2011 IFIP/IEEE International Symposium p 81-88
[6] Jammal M, Singh T, Shami A, Asal R and Li Y 2014 *Software defined networking: State of the art and research challenges* Computer Networks vol 72 p 74-98.
[7] Big Switch Networks 2012 *The open SDN Architecture* https://www.bigswitch.com/sites/default/files/sdn_overview.pdf
[8] Shin M K, Nam K H and Kim H J 2012 *Software-defined networking (SDN): A reference architecture and open APIs* (IEEE) In ICT Convergence (ICTC), 2012 Int. Conf. p 360-361
[9] Khasnabish B, Huang D, Bai X, Bellavista P, Martinez G and Antonopoulos N 2012 *Cloud Computing, Networking, and Services* Journal of Network and Systems Management vol 1 p 1-5.
[10] OpenStack 2013 *OpenStack Installation Guide for Ubuntu 14.04 (LTS)* http://docs.openstack.org/juno/install-guide/install/apt/content/

IOP Conf. Series: Materials Science and Engineering **121** (2016) 012004 doi:10.1088/1757-899X/121/1/012004

Kinetic Study on Pyrolysis of Oil Palm Frond

V S Y Soon[1], B L F Chin[1], A C R Lim[1]

[1] Department of Chemical Engineering, Faculty of Engineering and Science, Curtin University, Sarawak Malaysia CDT 250, 98009 Miri Sarawak, Malaysia.

Abstract. The pyrolysis of oil palm frond is studied using thermogravimetric analysis (TGA) equipment. The present study investigates the thermal degradation behaviour and determination of the kinetic parameters such as the activation energy (E_A) and pre-exponential factor (A) values of oil palm frond under pyrolysis condition. The kinetic data is produced based on first order rate of reaction. In this study, the experiments are conducted at different heating rates of 10, 20, 30, 40 and 50 K/min in the temperature range of 323-1173 K under non-isothermal condition. Argon gas is used as an inert gas to remove any entrapment of gases in the TGA equipment.

1. Introduction

Biomass is an attractive feedstock for providing potential source of hydrocarbons with almost zero net carbon dioxide (CO_2) emission to the atmosphere environment. The efficient utilization of biomass as a source of clean energy and chemicals has attracted much research attention in recent years amid fossil fuel energy as well as clean energy drive. According to International Energy Agency (IEA), biomass contributed approximately about 10% of the world total primary energy supply in year 2009 [1]. It is reported that an approximate 140 billion metric tons of biomass waste generated from agriculture activities alone worldwide for year 2009 [2]. The high quantity of biomass wastes show that the demand of crops and agricultural activities in the world is currently growing in an accelerating rate and it is expected to increase continuously in the future. Therefore, the abundant supply of biomass waste will eventually be an issue to the environment if it is left unattended.

In order to solve the abundant supply of biomass waste generated globally, pyrolysis process is selected among the thermochemical conversion methods in this present study to reduce the dumping of biomass waste and also providing a solution to produce useful chemicals and clean renewable energy to meet the increasing demand of the ever increasing human population. Biomass pyrolysis is a process where the biomass waste undergoes a thermal decomposition using a heat source in the absence of oxygen to produce energy and conversion of useful oils and char [3]. The few advantages of using pyrolysis compared to other thermochemical processes such as combustion, gasification and incineration are that it does not required the burning of oxygen and does not involved the release of large amount of harmful compounds into the atmosphere [4].

For the present work, palm oil frond is selected as the biomass for the pyrolysis process. One of the main reasons because palm oil wastes are the main biomass resources in ASEAN countries [5]. The two largest palm oil producers in the world are Malaysia and Indonesia which produces an estimated amount of 30 and 8.2 million tons respectively in year 2000 and said to increase yearly [5]. Among the palm oil residues, oil palm frond is said to produce the most abundant and it is reported to produce an estimated amount of 10.88 tons per hectares per year [6-8].

In biomass pyrolysis, thermal degradation behavior can be observed by analysing at the degradation of the three main components in biomass which are cellulose, hemicellulose, and lignin. The breakdown of these individual components occurred at different temperature. In theory, hemicellulose is the first component to decompose under pyrolysis process in the biomass. As hemicellulose is a component that is found abundantly in branches and having a random and amorphous structure, it provides the component to be decomposed easily into more evolved volatiles and lesser tar at lower temperature [9]. After the decomposition of hemicellulose, cellulose decomposes as it is a natural polymer which contained a long chain of linked smaller glucose molecules [10]. It is due to this characteristics of cellulose that gives it a strong strength structure. Thereafter, it is followed by the decomposition of lignin. Lignin is a big group of aromatic polymers where it is consists of several oxygen functional groups that results in requirement of higher heat to decompose lignin structures. Thus, the degree of temperature required to decompose biomass increases from hemicellulose, cellulose to lignin. However, thermal degradation behavior of these three components are different from each other and it is due to the component characteristics and structure of the component.

Few studies are reported in literature on the pyrolysis of biomass. Chin et al. [11] investigated on the pyrolysis of rubber seed shells using thermogravimetric analysis (TGA) equipment under non-isothermal condition in argon atmosphere to study the thermal degradation behavior and determination of kinetic parameters. Yang et al. [5] studied on the pyrolysis of palm oil wastes (palm shell, fiber, and empty fruit bunch) in a counter-current fixed bed to characterize the hydrogen rich gas products.

In the present study, TGA equipment is used to study the thermal degradation behaviour and kinetic analysis from the energy released from palm oil frond in pyrolysis process. The kinetic data is produced based on first order rate of reaction. In this study, the experiments are conducted at different heating rates of 10, 20, 30, 40 and 50 K/min in the temperature range of 323-1173 K under non-isothermal condition. Argon gas is used as an inert gas to remove any entrapment of gases in the TGA equipment.

2. Experimental

2.1. Material

Oil palm frond is obtained from a palm oil mill from West Malaysia. The characteristics of the oil palm frond is presented in Table 1.

Table 1. Characteristics of oil palm frond.

Proximate Analysis (weight%, dry basis)	
Moisture content	7.4
Volatile matter	72.5
Fixed carbon[a]	5.8
Ash	14.3
Ultimate Analysis (weight%, dry basis)	
Carbon	38.4
Hydrogen	5.5
Nitrogen	2.3
Sulphur	0.09
Oxygen	32.1
Other	21.6

Higher Heating Value (MJ/kg)	17.2

[a]By difference

2.2. Experimental Technique

The analytical instrument used in the present work is the TGA equipment. TGA is a technique where the change in mass of the material as a function of temperature or time is measured under controlled atmosphere. A TGA equipment includes a sample pan where it is supported by a precision balance. The pan laid in a furnace is either heated or cooled in the TGA equipment. The change in mass is monitored during the experiment and a purge gas which can be either inert or reactive controls the sample condition that passes through the sample in the TGA equipment.

In the present work, an approximately 5.0 mg of sample is filled in a 70 μL alumina crucible of the TGA equipment under inert atmosphere of argon. The argon gas is set at the temperature of 323 K with a flow rate of 100 mL/min and fed to the system for 20 minutes. The argon gas supplied is to get rid of the entrapment of other gases present in the system and prevent any unwanted oxidation on the sample in the pyrolysis zone. Subsequently, all of the samples are heated from the temperature of 323 K to 1173 K at its respective heating rates. During this process, the temperatures are kept unchanged for 10 minutes. During the heating, the TGA equipment measured the mass of the sample and furnace temperature. The experiment is conducted where the thermogravimetric curves are plotted at different heating rates of 10, 20, 30, 40 and 50 K/min within a range of temperature of 323 to 1173 K. The reproducibility of the experiments is checked for accuracy of the data obtained.

2.3. Determination of Kinetic Parameters

For pyrolysis process, the first order rate equation is represented as in Equation (1) [6]:

$$\text{Solid} \xrightarrow{k} \text{Volatiles} + \text{Char} \quad (1)$$

where volatiles is the total amount of the gas and tar meanwhile k is the constant rate of pyrolysis where the temperature dependence is written in Arrhenius equation as shown in Equation (2):

$$k = Ae^{(-E_A / RT)} \quad (2)$$

where E_A is the activation energy (kJ/mol), T is the absolute temperature (K), R is the gas constant (8.314J/kmol) and A is the pre-exponential factor (1/min).

The rate of conversion from biomass to volatile product is presented by the following expression shown in Equation (3):

$$\frac{dx}{dt} = k(T)f(x) \quad (3)$$

where x represent the degree of conversion of the pyrolysis process, t is the time, $k(T)$ is the rate constant and $f(x)$ is the reaction model.

Conversion, x is described as the normalized form of the weight loss of decomposed sample and it is represent as in Equation (4):

$$x = \frac{m_i - m_a}{m_i - m_f} \quad (4)$$

where m_i is the initial mass of the sample, m_a is the actual mass of the sample used, and m_f is the mass after the pyrolysis process.

Equations (3) and (4) are combined to give a fundamental expression of analytical methods to calculate the kinetic parameters on the basis of TGA results as indicated in Equation (5):

$$\frac{dx}{dt} = A \cdot f(x) \cdot e^{-E_A / RT} \tag{5}$$

Since the linear heating rate, β is expressed as follow:

$$\beta = dT / dt \tag{6}$$

Thus, the combined expression can be written as shown in Equation (7):

$$\frac{dx}{dt} = \frac{1}{\beta}(A \cdot f(x) \cdot e^{-E_A / RT}) \tag{7}$$

Rearranging the expression gives:

$$\frac{dx}{f(x)} = \frac{1}{\beta}(A \cdot f(x) \cdot e^{-E_A / RT})dT \tag{8}$$

By introducing $g(x) = \int_0^x \frac{dx}{f(x)}$, gives the following expression:

$$g(x) = \int_0^x \frac{dx}{f(x)} = \int_{T_I}^{T_F} \frac{1}{\beta}(A \cdot e^{-E_A / RT})dT \tag{9}$$

For first order reaction, $f(x) = 1\text{-}x$ meanwhile $g(x) = -\ln(1\text{-}x)$ provides:

$$-\ln(1 - x) = \int_0^x \frac{dx}{1-x} = \int_{T_I}^{T_F} \frac{1}{\beta}(A \cdot e^{-E_A / RT})dT \tag{10}$$

$$\ln[-\ln(1-x)] = \left(-\frac{E_A}{R}\right)\frac{1}{T} + \ln\left(\frac{ART^2}{\beta E_A}\right) \tag{11}$$

After plotting the linear graph of ln[-ln(1-x)] versus $1/T$, the value of E_A can be found by calculating the slope of the graph where it represents the $-E_A/R$ whereas the ln ($ART^2/\beta E_A$) represents the intercept-c. Therefore, the pre-exponential factor, A can be obtained by using the temperature as $T_a = (T_I + T_F)/2$ where T_I is the initial temperature of the main mass loss in every stage and T_F is the temperature of the final mass loss in every stage.

3. Results and Discussion

3.1. Thermogravimetric (TG) and Derivative Thermogravimetric (DTG) Analysis for Oil Palm Frond

The illustration of thermogravimetry analysis (TG) in terms of weight percentage (wt%) versus temperature (K) is plotted to show the thermal degradation behaviour of oil palm frond. The TG curves of oil palm frond at different heating rates from 10 K/min to 50 K/min is shown in Figure 1. In Figure 1, the curve for TG curves show similar trend for each heating rate in the studied range. From the values displayed in Table 2, it is shown that the residual weight in terms of wt% of the sample increases with the increased of heating rates. In addition, the temperature for each stage increases with heating rates. A similar behaviour is also observed as found in the literature [11].

In Figure 1, the TG curves illustrate a slow increment of mass loss for all heating rates studied where the temperature starts from room temperature, 298 K to a temperature of 500 K. At this phase, the moisture of the oil palm frond is eliminated and it is classified as the evaporation process. After the removal of water present in the oil palm frond, the curves of mass loss show a significant decrease of 500 K to 650 K in all heating rates and it shows a major decrease in the mass of the biomass. This further explained that the oil palm frond is facing main devolatilization phase where it released volatile matters and carbon. The mass loss continues to increase but at a slower rate from temperature of 650 K to 1050 K and remained relatively constant from temperature of 1050 K to 1200 K. After the experiment is conducted, the solid remaining is in the range of 8.0 to 15.9 wt% in the range of heating rate studied.

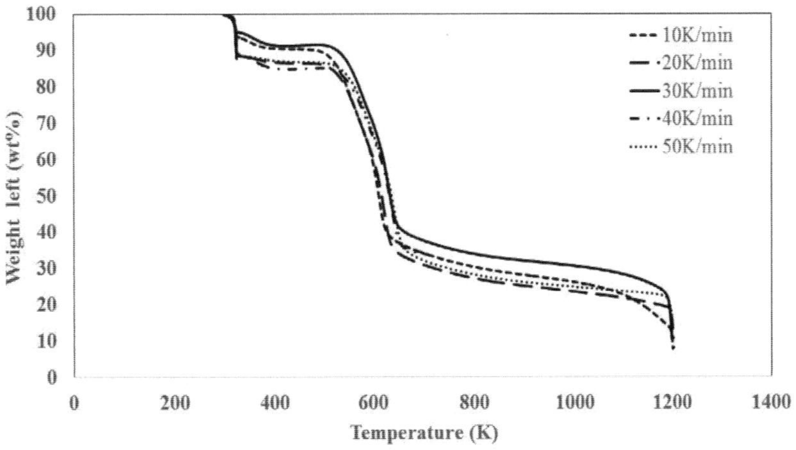

Figure 1. Thermogravimetric (TG) analysis for oil palm frond at different heating rates.

Table 2. Degradation temperature for oil palm frond.

Heating rate (K/min)	**First stage ($T_1/T_{max,1}/T_2$)**	**Second stage ($T_1/T_{max,2}/T_2$)**
10	468/559/571	571/614/641
20	471/573/585	585/622/663
30	480/581/598	598/631/687
40	488/591/600	600/639/695
50	495/594/604	604/642/704

Table 3. Amount of residue and maximum degradation rate values for oil palm frond.

Heating rate (K/min)	**Residue (wt%)**	**Maximum degradation rate (wt%/min)**
10	8.0	10.1
20	13.1	19.9
30	10.6	28.7
40	12.8	39.0
50	15.9	47.3

In Figure 2, the derivative thermogravimetry (DTG) analysis in terms of wt%/min versus temperature is presented. The kinetic scheme of DTG graph is illustrated with different heating rates. The DTG curve for pyrolysis of oil palm frond becomes more significant when heating rate increases from 10 to 50 K/min. This figure shows two peak exists which indicates two major decomposition stage occurring in the oil palm fronds as temperature increases from 323 to 1173 K. As the heating rate increased from 10 to 50 K/min, the DTG curves become more significant and distinct. A similar behaviour is seen in the results obtained from Rotliwala and Parikh [12]. The occurrence of this event is due to the slow decomposition of the lignin component in the oil palm frond sample. From this result, the occurrence of the two peaks in the DTG curve of the oil palm frond is heavily involved with the two groups of main reactions that includes the three main components found in biomass which are cellulose, hemicellulose and lignin. All the components present undergoes the process of decomposition at different temperature. The first peak from the DTG curve is associated with the decomposition of hemicellulose and some lignin meanwhile the second peak from the DTG curve is associated with the decomposition of the cellulose component in the biomass.

From the analysis on the temperature range of the curve, the thermal degradation carried out on the hemicellulose, cellulose and lignin in the present study are found very close range from other biomass pyrolysis found in literature [11-12]. At a low heating rate such as 10 K/min, the overlapping of the two peaks shows a smoother sloping baseline compared to other heating rates in the DTG curve is due to the slow decomposition of lignin component in the slow heating rate under a wider temperature range. The pure components of hemicellulose, cellulose and lignin are normally found in woody biomass and therefore, there is a slight variation on the degradation temperature range of the components from both aforementioned studies and current studies as the temperature range of thermal degradation is depending on the type of biomass used for the studies. The chemical characteristics of oil palm frond and woody biomass is different. Hence, this will results in different degradation temperature, however, a similar degradation behaviour is observed in both of these biomass. As observed in the DTG curves, there is an increasing trend of lateral shift in the maximum temperature degradation of the first peak ($T_{max,1}$) and second peak($T_{max,2}$) when the heating rate increases from 10 to 50 K/min in the oil palm frond. This occurrence is caused by the thermal lag which is defined as the difference in temperature of the sample environment controlled by the TGA equipment and the temperature of the sample [13]. Normally, thermal lag occurs in the decomposition of heat transfer at different heating rates, kinetics of the decomposition and heat conductivity property of the oil palm frond particle. The maximum degradation rate is recorded from the range of 10.1 to 47.3 wt%/min.

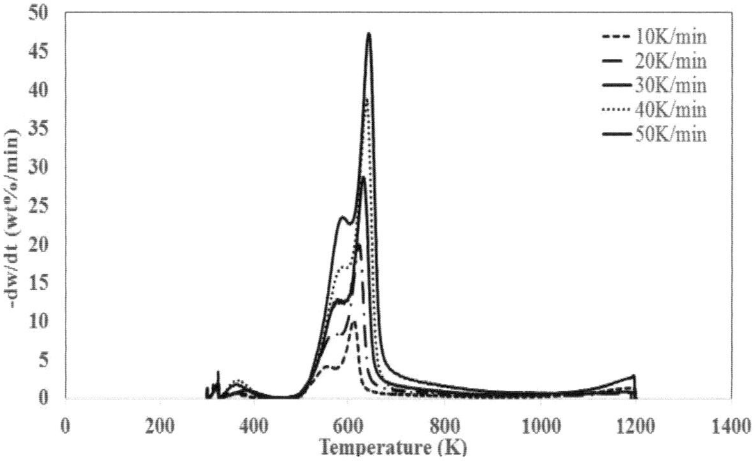

Figure 2. Derivative thermogrametric (DTG) analysis for oil palm frond at different heating rates.

3.2. Kinetic Analysis for Oil Palm Frond

The values for the kinetic parameters such as activation energy (E_A) and pre-exponential factor (A) are assumed to be in first order of reaction and are determined using an integral method are presented in Tables 4 and 5. Figure 3 displayed one of the exemplary plot of ln(-ln (1-x)) vs. $1/T$ using heating rate of 10 K/min. Similar trends are exhibit in different heating rates. The predicted E_A values ranged from 29.05 to 64.60 kJ/mol meanwhile the A values ranged from 45 to 176927 min^{-1} in the studied heating rates.

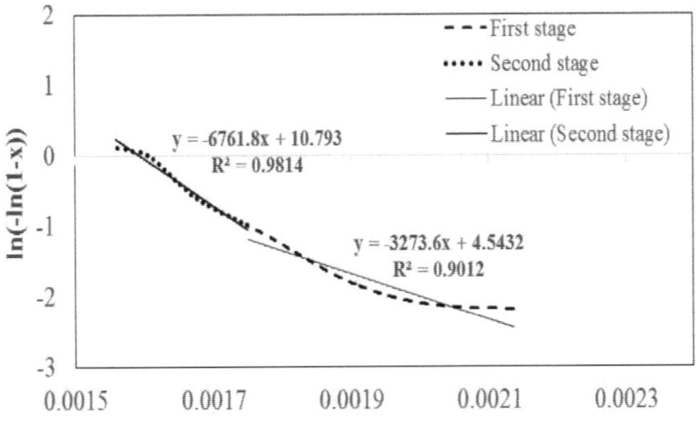

Figure 3. Arrhenius (ln(-ln(1-x)) versus 1/T) plot at 10 K/min for oil palm frond in pyrolysis process.

Table 4. Kinetic evaluation for pyrolysis of palm oil frond for first stage decomposition

Heating rate (K/min)	First stage		
	E_A (kJ/mol)	A (1/min)	R^2
10	33.77	55	0.9539
20	29.56	45	0.9506
30	38.11	316	0.9537
40	29.05	60	0.9543
50	32.92	196	0.9535

Table 5. Kinetic evaluation for pyrolysis of palm oil frond for second stage decomposition

Heating rate (K/min)	Second stage		
	E_A (kJ/mol)	A (1/min)	R^2
10	56.22	8828	0.9814
20	52.93	9200	0.9598
30	60.89	46826	0.9775
40	57.74	31578	0.9873
50	64.60	176927	0.9891

4. Conclusions

In this research study, oil palm frond is used as the biomass sample for pyrolysis process using the thermogravimetric analysis (TGA) equipment to study the thermal degradation behaviour and determine the kinetic parameters such as activation energy and pre-exponential factor values. The heating rate studied are from 10 to 50 K/min in the temperature range of 323-1173 K under non-isothermal condition. The temperature range for the first and second stage of degradation in the pyrolysis of oil palm fronds are 468-604 K and 571-704 K respectively. It is estimated that the activation energy and pre-exponential factor values is in the range of heating rate studied are 29.05 - 64.60 kJ/mol and 45-176928 min^{-1} respectively.

Acknowledgement

This work is carried out with financial support from Curtin University, Sarawak campus.

References

[1] International Energy Agency (IEA). http://iea.org/topics/bioenergy/
[2] United Nation Environment Programme 2009 *Conv. Waste Agri. Biomass into A Resourc.* (Japan).
[3] Jahirul M I, Rasul M G, Chowdhury A A and Ashwath N 2012 *Energies* **5** 4952-5001
[4] Grieco E M and Baldi, G 2012. *Waste Manag.* **32** 833-9
[5] Yang H, Yan R, Chen H, Lee D H, Liang and D T, Zheng, C 2006 *Fuel Process. Techno.* **87** 935-42
[6] Goyal H B, Seal D and Saxena, R C 2006. *Renew. Sust. Energy Rev.* **12** 504-17
[7] Saka S 2005. *Proc. of JSPS-VCC Natural Resour. & Energy Env. Seminar* (Japan).
[8] Singh G, Huan L K, Leng T and Kow D L 1999. Oil palm and the environment: a Malaysian perspectives. *Malaysia Oil Palm Growers Council* 41-53
[9] Sampath S S and Babu, B V 2006 *Proc. of the National Conf. on Env. Conserv.* (India: Rajasthan).
[10] Senese F M 2010. What is cellulose? http://antoine.frostburg.edu/chem/senese/101/consumer/faq/what-is-cellulose.shtml
[11] Chin, B L F, Yusup S, Al Shoaibi, A, Kannan P, Srinivasakannan, C and Sulaiman, S A 2014. *Energy Conv. Manag.* **87** 746-53
[12] Rotliwala Y and Parikh P 2011 *Korean J. Chem. Eng.* **28** 788-92
[13] Kersten S, Wang S Q, Prins W and Swaaij, W 2005 *Ind. Eng. Chem. Res.* **44** 8773-85

IOP Conf. Series: Materials Science and Engineering **121** (2016) 012005 doi:10.1088/1757-899X/121/1/012005

Performance of Silicon carbide whisker reinforced ceramic inserts on Inconel 718 in end milling process

M M Reddy and C X H Joshua
Faculty of Engineering and Science, Curtin University, Sarawak, Malaysia,
mohan.m@curtin.edu.my

Abstract. An experimental investigation is planned in order to study the machinability of Inconel 718 with silicon carbide whisker reinforced ceramic inserts in end milling process. The relationship between the cutting speed, feed rate, and depth of cut against the response factors are studied to show the level of significance of each parameter. The cutting parameters are optimized by using Taguchi method. Implementing analysis of variance, the parameter which influences the surface roughness the most is determined to be the cutting speed, followed by the feed rate and depth of cut. Meanwhile, the optimal cutting condition is determined to have high cutting speed, low feed rate, and high depth of cut in the range of selected parameters.

1. Introduction

Nickel-based superalloys have always been a significant usage in the aircraft and nuclear industry due to its excellent thermal resistance which maintains steady mechanical properties even at elevated temperature. However, being classified as a difficult-to-cut material, machining of nickel-based superalloy still remains as a challenge for various engineering experts in the industry due to the fact that it often generates numerous concerns and problems.

The constant demand to increase productivity and quality in machining operation has lead to the development of ceramic tools. Ceramic tools excel in high cutting speed and are much more suitable in high speed machining. In the industry field, the whisker-reinforced ceramic tools are mostly limited to semi-finishing and roughing operations due to their spontaneous failure and less predictable tool life [1]. Due to the reason that whisker reinforced ceramic are considerably expensive, it is fundamentally important to optimize the cutting parameters as finely as possible such that production requirements could be achieved.

Tool wear can be defined as a complicated phenomenon which is unavoidable and undesirable in every machining process. More importantly, tool wear has a strong presence in affecting production costs and also surface integrity of a machined component. The phenomenon where the chip and tool surface withstands high temperature and high relative velocity in addition to the chemical affinity of silicon carbide whiskers and Alloy 718 can be interpreted as leading to adhesion wear mechanism [2]. The resistance to depth-of-cut notch wear of the whisker-reinforced ceramic tool has been found to be greater than that of other Al_2O_3 based ceramic tools because of the toughening effect of the whiskers [4]. On the other hand, surface roughness is also a significant factor in machining as it indicates the surface quality of a machined part. Surface roughness is said to be related to tool wear in the sense of being directly affected by it whereby serious tool wear basically increases the roughness of the surface. Common cutting parameters include tool material, coating material, cutting speed, depth of cut, and feed rate. Being said, optimization of these parameters is very important to achieve minimum surface roughness and tool wears which relates to the aspect of cost reduction.

Design of experiments has the ability to predict the output with the minimum number of experiments by considering the experimental limits, specific conditions, and mathematical analysis [5]. The optimization of cutting parameters affecting surface roughness can be dealt with various methods such as Taguchi method and

IOP Conf. Series: Materials Science and Engineering **121** (2016) 012005 doi:10.1088/1757-899X/121/1/012005

multiple regression models. It is essential for every machining process to select optimal machining conditions due to the fact that it is a key factor in achieving higher machining efficiency. Taguchi method represents a method totally based on statistical design of experiments.

2. Taguchi Design

The Taguchi method represents an experimental design technique, which is very useful and effective in experimental investigation by significantly reduces the number of experimental runs. Hence, this technique is able to significantly reduce the time required for experimental investigation due to its ability to investigate multiple factors on performance and also the influence of individual factors [3]. In this study, the orthogonal array is formed with three factors while implementing three different levels. The three factors include process parameters of cutting speed, feed rate, and depth of cut. The L9 orthogonal array of this study mainly investigates the main effects of the process parameter towards surface roughness and flank wear, therefore the interaction between process parameters is neglected. Therefore, after the orthogonal array has been selected, the next step would be proceeding to run the experiment. Results of the cutting experiments are studied by using the S/N ratio and ANOVA. Based on the results of the analyses, optimal cutting parameters for minimum surface roughness and flank wear are obtained and verified.

3. Experimental Methodology

The experiments were carried out in end milling process without using coolant. End milling process is considered as the most common machining process in various industries due to its versatile and rapid machining. The work piece material that is employed in this research project is Inconel 718, a type of nickel-based super alloy where the work piece has a dimension of 200mm length, 100mm width, and 15.875mm thickness. Silicon carbide whisker reinforced ceramic inserts collected from SSANGYONG company graded SW500. This grade of ceramic insert is a newly developed ceramic cutting tool where the company state that it has excellent flank and notch wear resistance in high speed cutting especially in machining high temperature alloy and high nickel alloy.

Table 1: Range of cutting parameter values

Parameter	Unit	Level of parameter		
		Level 1	Level 2	Level 3
Cutting speed	m/min	200	250	300
Feed rate	mm/rev	0.05	0.07	0.09
Depth of cut	mm	0.50	0.70	0.80

Table 2: Parameter level for L9 orthogonal array

	Process parameters		
Experiment Run	Cutting speed	Feed rate	Depth of cut
1	1	1	1
2	1	2	2
3	1	3	3
4	2	1	2
5	2	2	3
6	2	3	1
7	3	1	3
8	3	2	1
9	3	3	2

The process parameters range, selection of levels are listed in Table 1, which is decided based on the recommendation from the tool suppliers and literature. The Taguchi L9 orthogonal array defines the 9 tests for three cutting parameters are listed in Table 1. The surface roughness of the work piece is measured by using Mitutoyo SJ-301 surface roughness tester.

IOP Conf. Series: Materials Science and Engineering **121** (2016) 012005 doi:10.1088/1757-899X/121/1/012005

4. Results and Discussion

4.1. S/N ratio ANOVA results

The signal-to-noise ratio is determined to identify optimal control factor settings which make the process robust. S/N ratio transformation preferred is smaller the better characteristics, which follows the formula of equation 1 [6],

$$S/N = -10 \log \frac{1}{n} (\sum y^2). \tag{1}$$

Based on the analysis of the S/N ratio, the greater S/N ratio corresponds to the smaller variance of the output characteristic around the desired value. Hence, as illustrated in the S/N response graph above, the highest S/N ratio for surface roughness are obtained at cutting speed level 3 (300m/min), feed rate level 1 (0.05mm/rev), and depth of cut level 3 (0.8mm). Therefore, the optimal cutting parameters to obtain better surface roughness can be determined to have high cutting speed, low feed rate, and high depth of cut.

Table 3: Surface roughness experiment results

Experiment run	Cutting speed (m/min)	Feed rate (mm/rev)	Depth of cut (mm)	Surface roughness R_a (μm)	S/N ratio
1	200	0.05	0.50	1.79	-5.06
2	200	0.07	0.70	1.67	-4.45
3	200	0.09	0.80	1.78	-5.01
4	250	0.05	0.70	1.74	-4.81
5	250	0.07	0.80	1.80	-5.11
6	250	0.09	0.50	1.68	-4.51
7	300	0.05	0.80	1.40	-2.92
8	300	0.07	0.50	1.72	-4.71
9	300	0.09	0.70	1.75	-4.86
Total					-41.44

Table 4: Average S/N ratio for surface roughness

Average S/N ratio for Surface roughness			
Parameter	Level 1	Level 2	Level 3
Cutting speed	-4.84	-4.81	-4.16
Feed rate	-4.26	-4.76	-4.79
Depth of cut	-4.76	-4.71	-4.35

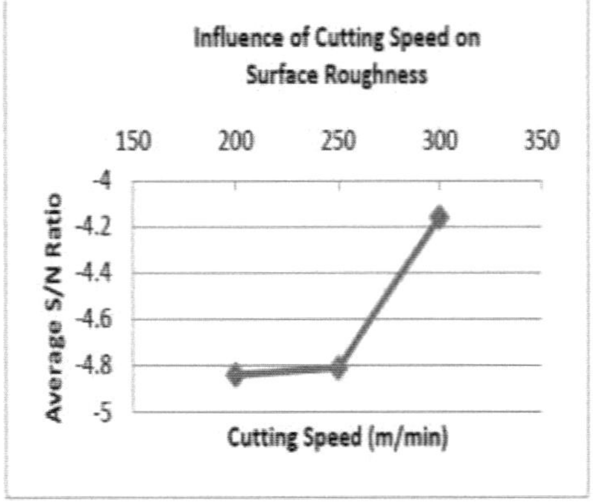

Figure 1: Influence of cutting speed

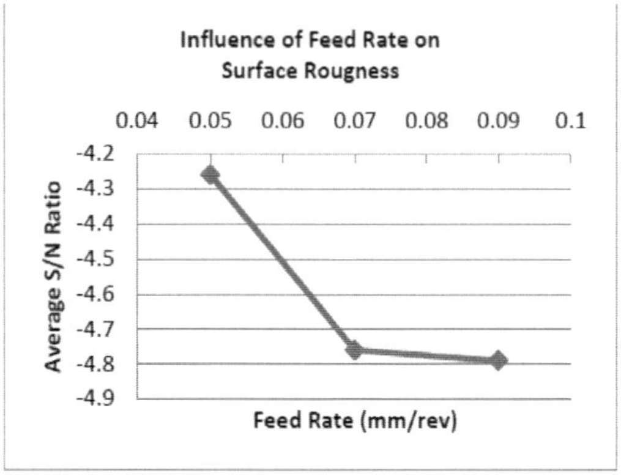

Figure 2: Influence of feed rate

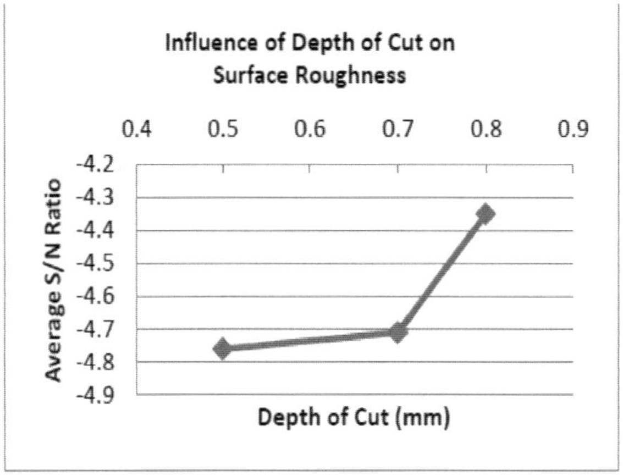

Figure 3: Influence of depth of cut

Table 5 shows the analysis of variance (ANOVA) results with 95% confidence level in order to determine the significant parameter. Through the ANOVA analysis, the contribution for each parameter is identified where cutting speed contributes 24.25%, feed rate having 14.53%, and depth of cut having 8.39% contribution factor. This indicates that the cutting speed is the most significant parameter in influencing the surface roughness followed by the feed rate and lastly the depth of cut.

Table 5: Results for analysis of variance

Analysis of variance for surface roughness				
Parameter	Degree of freedom	Sum of squares	Mean square	Contribution (%)
Cutting speed	2	0.8770	0.4385	24.25
Feed rate	2	0.5256	0.2628	14.53
Depth of Cut	2	0.3033	0.1517	8.39
Error	2	1.9105	0.9553	52.83
Total	8	3.6164		100

4.2. Tool wear results

Wear patterns observed in cutting tools results from the combined effect of different wear mechanisms involved in sliding friction occurring at the contact interface between tool, chip and machined surface. In the case of machining nickel-based super alloys, the most known tool wear pattern would be flank wear and notch wear. The main concern and focus on the tool wear in this experiment would be to determine the results of flank wear which is generally caused by the fracture in ultra- high speed cutting process. Basically, all of the figures of tool wear shown in Figure 4, do not seem to illustrate a significant difference as all nine set of cutting tools display similar tool wear and pattern where tool fracture had occurred.

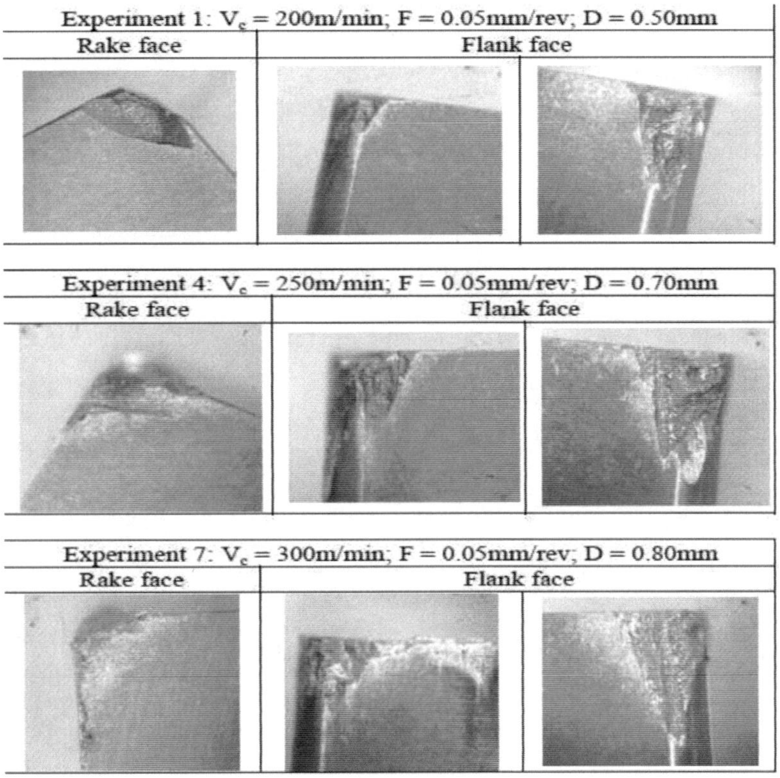

Figure 4: Tool wear pattern

Tool fracture would occur with the increase in the volume of metal removal, as the tool is damaged progressively by the alternating mechanical stress and thermal stress. This is most probably due to the major disadvantages of ceramic tools where they have low fracture toughness while being low resistance to mechanical shock as well. Furthermore, by observing the respective fracture, the fracture can be said to be too serious for further machining which indicates that the tool life of the cutting tool had been reached.

5. Conclusion

There are several conclusions which can be drawn for this research work. Firstly, the S/N ratio analysis identifies that the optimal cutting condition for the experiment is by having high cutting speed, low feed rate, and high depth of cut. Analysis of Variance (ANOVA) proves that cutting speed is the factor that influences the surface roughness the most with a contributing factor of 24.25%, followed by feed rate of 14.53% and depth of cut with 8.39%. Although the main focus on tool wear in this project is stated to be on flank wear, however the results for tool wear is too extreme and is considered as tool failure whereby the tool fracture had been too serious. Therefore, it can be concluded that the SSANGYONG SW500 graded whisker reinforced ceramic cutting insert is suit only for rough machining of Inconel 718 due to extreme tool wear and unfavorable surface roughness.

IOP Conf. Series: Materials Science and Engineering **121** (2016) 012005 doi:10.1088/1757-899X/121/1/012005

REFERENCES

[1] Bushlya, Volodymyr, Jinming Z, Pajazit A, and Eric J S., 2012. "Performance and Wear Mechanisms of Whisker-Reinforced Alumina, Coated and Uncoated Pcbn Tools When High-Speed Turning Aged Inconel 718." *The International Journal of Advanced Manufacturing Technology* 66 (9-12): 2013-2021.

[2] Bushlya, Volodymyr, Jinming Z, Pajazit A, and Eric J S., 2013. "Wear Mechanisms of Silicon Carbide-Whisker-Reinforced Alumina (Al2o3–Sicw) Cutting Tools When High-Speed Machining Aged Alloy 718." *The International Journal of Advanced Manufacturing Technology* 68 (5-8): 1083-1093.J. Clerk Maxwell, A Treatise on Electricity and Magnetism, 3rd ed., vol. 2. Oxford: Clarendon, 1892, pp.68–73.

[3] Ghani J. A., Choudhury I A. and Hassan H H, 2004. "Application of Taguchi Method in the Optimization of End Milling Parameters." *Journal of Materials Processing Technology* 145 (1): 84-92.

[4] Jun, Zhao, Deng J, Zhang J, and Xing A., 1997. "Failure Mechanisms of a Whisker-Reinforced Ceramic Tool When Machining Nickel-Based Alloys." *Wear* 208 (1–2): 220-225.

[5] Reddy M M, Gorin A, Khaled A H., and.Sujan D., 2014. "Machinability of Aluminum Nitride Ceramic Using TiAlN and TiN Coated Carbide Tool Insert" *Materials Science Forum,* 773 : 437-447.

[6] Siddiquee, Arshad N, Zahid A K, Pankul G, Kumar M, Agarwal G, and Noor Z K. 2014. "Optimization of Deep Drilling Process Parameters of Aisi 321 Steel Using Taguchi Method." *Procedia Materials Science* 6: 1217-1225.

IOP Conf. Series: Materials Science and Engineering **121** (2016) 012006 doi:10.1088/1757-899X/121/1/012006

Structural and electrochemical characterization of $BaCe_{0.7}Zr_{0.2}Y_{0.05}Zn_{0.05}O_3$ as an electrolyte for SOFC-H

Ahmed Afif [1,*], Nikdalila Radenahmad[1], Chee Ming Lim[2], Quintin Cheok[1], Md. Aminul Islam[3], Seikh Mohammad Habibur Rahman[4], Abul Kalam Azad[1].

[1]Universiti Brunei Darussalam, Faculty of Integrated Technologies, Jalan Tunku Link, Gadong BE 1410, Brunei Darussalam.
[2]Universiti Brunei Darussalam, Center for Advanced Materials and Energy Sciences,Jalan Tunku Link, Gadong BE1410, Brunei Darussalam.
[3]Universiti Brunei Darussalam, Faculty of Sciences, Jalan Tunku Link, Gadong BE1410, Brunei Darussalam.
[4]Chalmers University of Technology, Department of Chemistry and Chemical Engineering, Goteborg SE 41296, Sweden.

*Corresponding author: Email: 14m0506@ubd.edu.bn, Tel.: +6738953255

Abstract. As a potential electrolyte for proton-conducting solid oxide fuel cells (SOFC-Hs) and to get better protonic conductivity and stability, zinc doped BCZY material has been found to be promising. In this study, we report a new composition of proton conductors $BaCe_{0.7}Zr_{0.2}Y_{0.05}Zn_{0.05}O_3$ (BCZYZn5) which was investigated using XRD, SEM and conductivity measurements. Rietveld refinement of the XRD data revel a cubic perovskite structure with Pm-3m space group. $BaCe_{0.7}Zr_{0.2}Y_{0.05}Zn_{0.05}O_3$ shows cell parameter a = 4.3452(9) Å. Scanning electron microscopy images shows that the grain sizes are large and compact which gives the sample high density and good protonic conductivity. The total conductivity in wet atmosphere is significantly higher than that of dry condition and the conductivity was found to be 0.276 x 10^{-3} Scm^{-1} and 0.204 x 10^{-3} Scm^{-1} at 600°C in wet and dry Ar, respectively. This study indicated that perovskite electrolyte BCZYZn5 is a promising material for the next generation intermediate temperature solid oxide fuel cells (IT-SOFCs).

1. Introduction

These invention of Solid Oxide Fuel cells (SOFC) brings a great advantage in renewable and sustainable energy system for future generation. This is the most efficient device among the energy technology invented so far [1]. In recent years proton conducting electrolytes brings a great advantage in SOFCs technologies to operate in intermediate temperatures. They have a growing interests in proton-conducting oxide electrolytes for SOFCs and a wide range of technological applications in fuel cells, batteries, gas sensors, hydrogenation or dehydrogenation of hydrocarbon electrolysers [2-4]. A lots of perovskite type oxides show high proton conductivity in H_2 and H_2O containing atmospheres at relatively low operating temperatures (400-700°C), low activation energy and high efficiency [5, 6].
Finding the best proton conducting and proper compromise perovskite material between conductivity and chemical stability is a great challenge. To get highly densified materials at low sintering temperature

for proton conducting electrolyte is also a major challenges. $BaCeO_3$-based materials may exhibit mixed ionic (oxide ions and protons) conduction [7, 8]. Doped barium cerates ($BaCeO_3$) possess high proton conductivity and good sintering behavior [9, 10], though they are unstable in CO_2 and steam, destroying the perovskite structure [11], which is essential to the maintenance of high proton conductivity. On the other hand doped zirconates which offer low proton conductivity and poor sinter ability but high chemical stability in CO_2 and H_2O atmospheres [12-16]. In combining these two doped, an apparent optimized material could be obtained. In recent reports the solid solutions of $BaCeO_3$ and $BaZrO_3$ combined the high proton conductivity of barium cerate with the good chemical stability of barium zirconate [17, 18], despite the fact that the sintering temperature was still very high [19]. Recently it was reported that the better chemical stability obtained by partial substitution of Zr^{+4} cations into Ce^{+4} cations as an electrolyte material [20, 21]. The introduction of small amount of Zn at B-site into the structure allows a reduction in high sintering temperatures and a remarkable improvement in the stability, relative density and conductivity [22]. As a potential high performance proton conductor BCZYZ has been offering the benefits of both stability and increased proton conductivity [23]. These oxides also exhibit mixed proton and oxide-ion conductivity [24]upon exposure to humid atmosphere. This new material for proton conducting electrolyte offers to minimize those challenges.
In this study we report a new proton conducting perovskite series $BaCe_{0.7}Zr_{0.2}Y_{0.05}Zn_{0.05}O_3$ to overcome those major challenges. We have studied the thermal and electrochemical properties of the series using XRD, SEM and conductivity measurements.

2. Experimental

The zinc doped BCZY powder materials were prepared by using solid state reaction method. Stoichiometric amount of $BaCO_3$, CeO_2, ZrO_2, Y_2O_3 and ZnO were mixed by ball milling with zirconia balls and acetone to ensure thorough mixing for 24 h. The finely ground dried materials were fired at 1000°C for 8 h with a fixed heating rate of 5°C min^{-1} and subsequently ground and palletized using 13mm diameter die and Kennedy Hydraulic Bench Press under pressure of 202 MPa (3 ton) and sintered at 1200°C in air for 18 h with a fixed heating rate of 5°C min^{-1}. The final sintering temperature was 1400°C in air for 22 h with a fixed heating rate of 5°C min^{-1} using Nabertherm box furnace.

The phase structure, purity, identity and homogeneity of the mixed oxide was examined by at ambient temperature using Shimadzu-7000 diffractometer ($CuK\alpha_1$=1.5406 Å) in the 2θ range from 10° to 80°. The data were collected with a 0.01° step size and a count time of 60 s/step. The obtained data were refined by the Rietveld method using the FullProf software [25]. The microscopic features of the prepared electrolytes were characterized using sub-nanometer resolution Scanning Electron Microscope (JSM-7610F). The SEM data were collected in atmospherically-isolated chamber.

To measure the deformation of the sample materials under non-oscillating stress against time or temperature, with programmed temperature using SETARAM Instrumentation-SETSYS Evolution TMA S60/28682-LCT10414-3. For the BCZYZn5 material, the thermal expansion coefficients properties are measured by thermomechanical analysis. To determine both the size of particles and their state of distribution of BCZYZn5 material, Horiba Particle Size Distribution Analyzer LA-920 and Horiba Reservoir Unit LY-201 were used in wet mode.

The conductivity of the sample was analyzed by using Impedance Spectrometry. A Solartron 1260 frequency response analyzer connected to a ProboStat (NorECs, Norway) conductivity cell was used to measure impedance over the frequency range 6 MHz to 1Hz and the applied sine wave amplitude was 1 V rms. a sintered pellet of the as-prepared material, 13 mm in diameter, and with approximately 0.5 cm^2 platinum paste electrodes were used. Impedance data were initially collected on cooling from 1000 to 150°C (cooling cycle) in 50°C steps under flowing dry and wet Ar. The gas was passed through two beds of P_2O_5 desiccant before entering the conductivity cell and is called the 'Dry Ar'. Whereas, when Ar gas was flowed through water at 22°C ($p(H_2O)$ = 0.026 atm) and the measurement is called 'Wet Ar'. Each time, sufficient time was allocated at each temperature to ensure equilibrium before spectra was

IOP Conf. Series: Materials Science and Engineering **121** (2016) 012006 doi:10.1088/1757-899X/121/1/012006

recorded. The least squares refinement program Z-View (Scribner Associates Inc.) was used to fit the obtained impedance data.

3. Results and discussions

3.1. X-ray Diffraction Analysis

Rietveld refinement of XRD data shows the $BaCe_{0.7}Zr_{0.2}Y_{0.05}Zn_{0.05}O_3$ material as cubic symmetry in the space group Pm-3m. Table 1 shows the unit cell parameter is **a** = 4.3452(9) Å.

Table 1. Summary of result obtained from Rietveld analysis of X-ray diffraction data for BCZYZn5

Sample	Space Group	Lattice Parameter (Å)			Chi² (χ^2)
		a	b	c	
$BaCe_{0.7}Zr_{0.2}Y_{0.05}Zn_{0.05}O_3$	Pm-3m	4.3452(9)	4.3452(9)	4.3452(9)	1.36

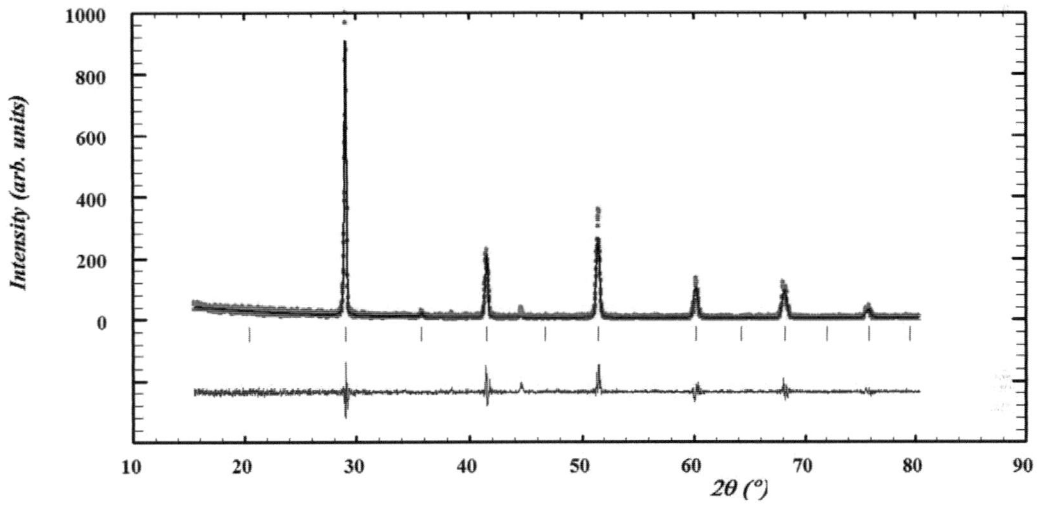

Figure 1. Rietveld analysis of X-ray diffraction data for BCZYZn5 as Cubic.

Figure 1 shows the Rietveld refinement of XRD of BCZYZn5 in cubic symmetry. The refinement converged quickly with significant improvements to the fit achieved when oxygen occupancies were allowed to vary from full occupancy. The occupancies of A and B site components were also checked and found to be fully occupied. The refinement of oxygen sites indicate the overall oxygen contents of O2.96 which is very close to the value required to conserve charge neutrality. The B-site occupancies were found as the expected stoichiometry within the limit of standard deviation. Decreasing Zr concentration leads to an increase of the lattice parameters. This increase of parameters reflects the substitution of small Zr4+ for Y3+-site. These results suggest that these compositions could be good approach as electrolyte materials in SOFC-H.

3.2. Scanning Electron Microscope Analysis

To observe the microstructures of the sample series, scanning electron microscopy was applied. Figure 2 a) shows the SEM top-view of surface morphology of the as-prepared BCZYZn5 materials. The surface was smooth and free of cracks. Thus, this compositions indicated non porosity, no trace of liquid or secondary phases were found at the grain boundary region in any of the samples investigated. It also shows that the grain sizes are larger and compact full which gives high density. Figure 2 b) shows the similar work but sample composition was different [26]. Comparing this two figures that for Figure 2 a), the grain size is much bigger than 2 b). For BCZYZn5 the small spherical phase is approximately 1μm. The large grain size gives less grain boundary resistance also. The SEM data were collected with a voltage of 5 kV and with magnification of 3000 times. These results suggest that the introducing of Zn leads to a compactness of the grain size.

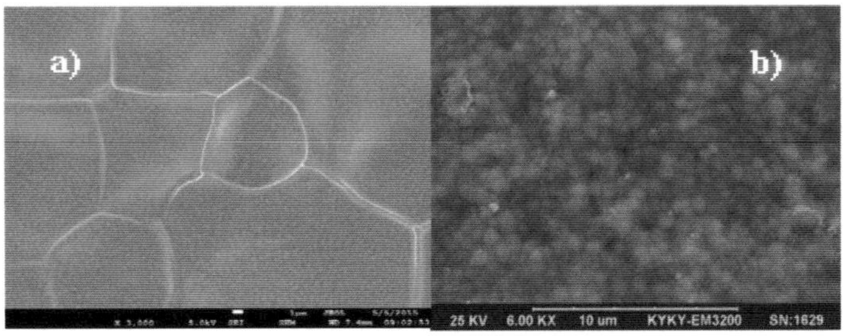

Figure 2. a) SEM analysis of $BaCe_{0.7}Zr_{0.2}Y_{0.05}Zn_{0.05}O_3$, b) SEM image of $BaCe_{0.5}Zr_{0.3}Y_{0.16}Zn_{0.04}O_{3-\delta}$ [26].

3.3. Thermo-Mechanical Analysis

The result of thermal expansion curves for $BaCe_{0.7}Zr_{0.2}Y_{0.05}Zn_{0.05}O_3$ oxide materials, shows the lowest of all temperature ranges. In Figure 3, the thermal expansion coefficient (TEC) increases with the increase of Yttriam % and also with the temperature. The TEC is also increased with the decrease of Zirconium %. To analyze the thermomechanical properties of the composition, the TEC is 10.919×10^{-6}/°C at 898.92°C.

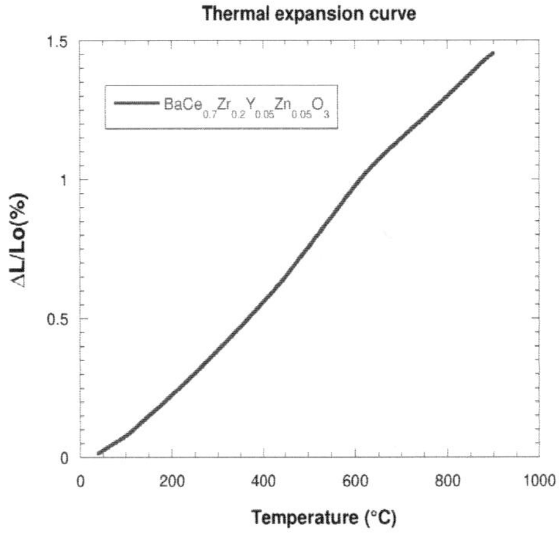

Figure 3. Thermal expansion curve for $BaCe_{0.7}Zr_{0.2}Y_{0.05}Zn_{0.05}O_3$

3.4. Particle Size Measurement

The The particle size distribution for the metal oxides used in this investigation was calculated in powder after sintering at 1400°C. The metal-oxide powders were previously sonicated during 4 h at 60°C to obtain a homogeneous particles dope. Table 2 shows that, as the increase of increasing of Y concentration with Zr leads to a decrease of the particle size. The addition of metal oxide particles not only increases water diffusion into the growing membrane due to its higher hydrophilic nature but also affects the interaction between polymer and solvent molecules by the hindrance effect of the particles. From figure 4 and, it was indicated that the quantity (%) of the sample varies with the diameter of particle size and the diameter range of the BCZYZn5 material remain within 0-2.5 μm.

Table 2. Particle size analysis: with standard deviation

Sample	Quantity (%)	Diameter (μm)	Standard Deviation (S.D)(μm)
$BaCe_{0.7}Zr_{0.2}Y_{0.05}Zn_{0.05}O_3$	6.856	1.981	0.9329

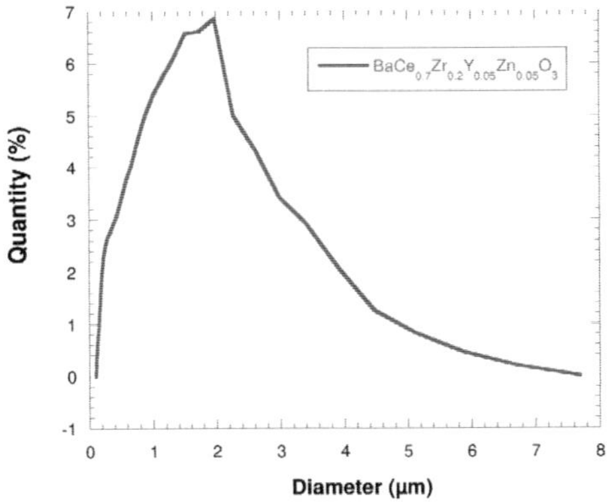

Figure 4. Quantity vs Diameter curve of $BaCe_{0.7}Zr_{0.2}Y_{0.05}Zn_{0.05}O_3$ for particle size.

3.5. Impedance Spectroscopy

The total and bulk conductivity increases along with the concentration of Y^{3+} dopant. This trivalent, dopant creates more vacant oxygen sites and in turn increases the proton concentration. Figure 5 shows the Arrhenius plot of the sample in hydrate (under dry Ar) and as-prepared (under wet Ar) form. The bulk conductivity in wet and dry Ar atmosphere was 0.00136 Scm^{-1} and 0.000946 Scm^{-1} at 600°C and, 0.004242 Scm^{-1} and 0.00331 Scm^{-1} at 900°C, respectively. Similarly the total conductivity in wet and dry Ar atmosphere was 0.000276 Scm^{-1} and 0.000204 Scm^{-1} at 600°C, and 0.003014 Scm^{-1} and 0.00218 Scm^{-1} at 900°C respectively. Figure 6 shows that the conductivity increases with temperature as well as wet atmosphere. The increase of total conductivity is smoother than the bulk conductivity. Slodczyk et al. reported that the BCZYZ ceramic – offering an high conductivity and stability as a potential candidate for use as an electrolytic membrane which was suited to the surface and bulk analysis [23]. In the present work, we have got higher conductivity in BCZYZn5 in comparison to BCZYZ reported by Slodczyk et al.

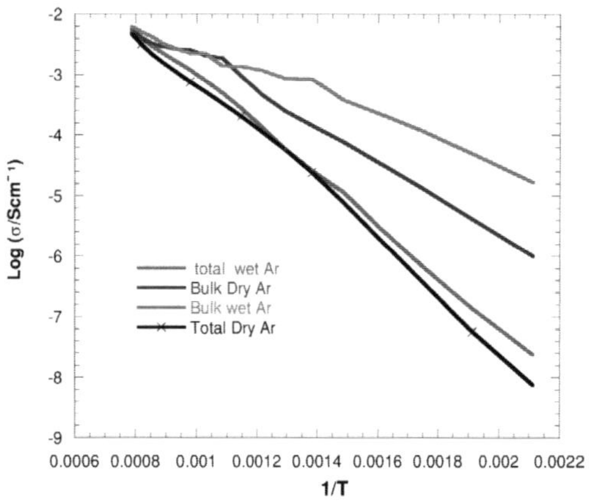

Figure 5. Arrhenius plot of the sample BCZYZn5 in dry and wet Ar. Bulk and total conductivity is also shown.

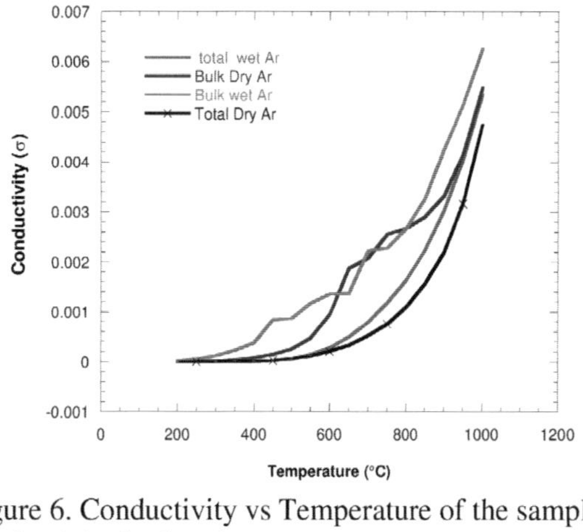

Figure 6. Conductivity vs Temperature of the sample BCZYZn5 in dry and wet Ar. Bulk and total conductivity is also shown.

4. Conclusions

The In this study, $BaCe_{0.7}Zr_{0.2}Y_{0.05}Zn_{0.05}O_3$ proton conducting electrolyte was successfully fabricated. Rietveld analysis of the XRD data indicates that the samples crystallize in cubic symmetry. The SEM analysis shows the electrolyte materials are non-porous and highly dense. The electrical measurements performed in different atmospheres have shown that adding Zn was beneficial in terms of increased conductivity values. For BCZYZn5, Bulk conductivity reaches 1.36×10^{-02} at 600°C in wet condition with an activation energy of 0.58 eV. To get high protonic conductivity and high stability zinc doped BCZYZn5 material has been found to be promising as a potential electrolyte for proton-conducting SOFCs.

5. Acknowledgment
The authors, AA and NR, are grateful to Ministry of Foreign Affairs and Trade (MOFAT), Brunei Darussalam, for providing scholarship to perform this research. We acknowledge the support from the Center for Advanced Materials and Energy Sciences (CAMES), Universiti Brunei Darussalam through Brunei Research Council grant no. S & T17.

6. References
[1] Badwal S, Giddey S, Kulkarni A and Munnings C 2014 *J. Aust. Cer. Soc* **50** 23-37
[2] Iwahara H 1995 *Solid State Ionics* **77** 289-298
[3] Norby T 1999 *Solid State Ionics* **125** 1-11
[4] Colomban P 1992 *Cambridge University Press* **2**
[5] Bi L, Tao Z, Liu C, Sun W, Wang H and Liu W 2009 *Journal of Membrane Science* **336** 1-6
[6] Guo Y, Ran R and Shao Z 2010 *International journal of hydrogen energy* **35** 5611-5620
[7] Iwahara H 1992 *Solid State Ionics* **52** 99-104
[8] Azad A K and Irvine J T 2008 *Solid State Ionics* **179** 678-682
[9] Hibino T, Hashimoto A, Suzuki M and Sano M 2001 *The Journal of Physical Chemistry B***105** 11399-11401
[10] Azad A K, Kruth A and Irvine J T 2014 *International journal of hydrogen energy* **39** 12804-12811
[11] Matsumoto H, Kawasaki Y, Ito N, Enoki M and Ishihara T 2007 *Electrochemical and Solid-State Letters* **10** B77-B80
[12] Chen F, Sørensen O T, Meng G and Peng D 1997 *Journal of Materials Chemistry* **7** 481-485
[13] Fabbri E, Epifanio A D, Bartolomeo E D, Licoccia S and Traversa E 2008 *Solid State Ionics* **179** 558
[14] Bhide S V and Virkar A V 1999 *Journal of The Electrochemical Society* **146** 2038-2044
[15] Babilo P, Uda T and Haile S M 2007 *Journal of materials research* **22** 1322-1330
[16] Ahmed I, Karlsson M, Eriksson S G, Ahlberg E, Knee C S and Larsson K 2008 *Journal of the American Ceramic Society* **91** 3039-3044
[17] Azad A K, Savaniu C, Tao S, Duval S, Holtappels P and Ibberson R M 2008 *Journal of Materials Chemistry* **18** 3414-3418
[18] Kreuer K 1999 *Solid State Ionics* **125** 285-302
[19] Bohn H G and Schober T 2000 *Journal of the American Ceramic Society* **83** 768-772
[20] Ryu K H and Haile S M 1999 *Solid State Ionics* **125** 355-367
[21] Katahira K, Kohchi Y, Shimura T and Iwahara H 2000 *Solid State Ionics* **138** 91-98
[22] Tao S and Irvine J T 2006 *Advanced Materials* **18** 1581-1584
[23] Slodczyk A, Sharp M D, Upasen S, Colomban P and Kilner J A 2014 *Solid State Ionics* **262** 870-874
[24] Iwahara H, Uchida H, Ono K and Ogaki K 1988 *Journal of the Electrochemical Society* **135** 529-533
[25] Rodríguez J and Carvajal 1993 *Physica B: Condensed Matter* **192** 55-69
[26] Lin B, Dong Y, Wang S, Fang D, Ding H and Zhang X 2009 *Journal of Alloys and Compounds* **478** 590-593

Biodiesel Production using Heterogeneous Catalyst in CSTR: Sensitivity Analysis and Optimization

L S Keong[1], D S Patle[2], S R Shukor[3], Z Ahmad[4]
[1,3,4]School of Chemical Engineering, Engineering campus, Universiti Sains Malaysia (USM), 14300, Nibong Tebal, Penang, Malaysia
[2]School of Mechanical and Building Sciences (Chemical Department), VIT Vellore, 632 014, Tamil Nadu, India

Abstract. Biodiesel as a renewable fuel has emerged as a potential replacement for petroleum-based diesels. Heterogeneous catalyst has become the focus of researches in biodiesel production with the intention to overcome problems associated with homogeneous catalyzed processes. The simulation of heterogeneous catalyzed biodiesel production has not been thoroughly studied. Hence, a simulation of carbon-based solid acid catalyzed biodiesel production from waste oil with high FFA content (50 weight%) was developed in the present work to study the feasibility and potential of the simulated process. The simulated process produces biodiesel through simultaneous transesterification and esterification with the consideration of reaction kinetics. The developed simulation is feasible and capable to produce 2.81kmol/hr of FAME meeting the international standard (EN 14214). Yields of 68.61% and 97.19% are achieved for transesterification and esterification respectively. Sensitivity analyses of FFA composition in waste oil, methanol to oil ratio, reactor pressure and temperature towards FAME yield from both reactions were carried out. Optimization of reactor temperature was done to maximize FAME products.

1. Introduction

Biodiesel has emerged as one of the potential alternative fuels to replace petroleum-based fuels as its properties are similar to that of petroleum-based diesel [1]. Biodiesel has an edge over petroleum-based fuels in view of its sustainability as it is biodegradable, non toxic and biodiesel produces less pollutants after it is burned. There are no modifications required to the current automobile engine in order for the engine to run on biodiesel blend. Biodiesel produced from waste cooking oil can be used in engines without modification for up to 30% blend [2]. Biodiesel consists of a mixture of fatty acid methyl esters (FAMEs) which can be produced through transesterification of vegetables fats or animal oils with methanol. The reaction can be represented as shown in Figure 1.Biodiesel production through transesterification is usually carried out by homogeneous alkaline catalyst like NaOH and KOH. Although ethanol can be used for biodiesel production, the use of methanol in biodiesel production is more common and preferable due to the cheaper price and the large available feedstock of methanol. Esterification, required when FFA content is more, on the other hand is usually catalysed by a homogeneous acid catalyst like sulfuric acid [1,3]. Other approaches for biodiesel production are: use of heterogeneous catalyst, supercritical methanol and enzymes in the reactions [4]. These methods are still under research and development, and have become the focus of research in recent years. Many

[4] To whom any correspondence should be addressed.

studies have considered heterogeneous catalyst in order to get rid of the problems faced in homogeneous catalyst. Heterogeneous catalysts ease the separation process as heterogeneous catalyst removal is easier that homogeneous catalyst removal. Heterogeneous catalyzed biodiesel production required less process unit compared to the homogeneously catalyzed process [3]. Thus, the process equipment cost is generally lower compared to homogeneously catalyzed process. Thus, in this study, the heterogeneously catalyzed biodiesel production process is simulated. Later, sensitivity analysis and optimization are carried out to determine the effect of operating parameters and find optimal parameters.

Figure 1. Transesterification chemical equation

2. Biodiesel production using heterogeneous catalyst

The process is developed based on the works of West et al [5]. The same production capacity (800 metric tonnes per annum) is assumed in this process. However, instead of a packed bed reactor, a CSTR is used for the simultaneous reactions of esterification of FFA and transesterification of triglycerides. The reactions and the composition of waste oil are based on the reactions reported by Shu et al [6]. Shu et al [6] used a carbon based solid acid catalyst from petroleum asphalt for the simultaneous reactions. The temperature used for the reaction is chosen as 200°C which is the mean temperature of the temperature range in the study of Shu et al [6]. High pressure is commonly used for the heterogeneous biodiesel production. This pressure is chosen as this is higher than the methanol vapour pressure at the reaction temperature [7]. Thus, 40 bar is used for the present study. Figure 2 shows the schematic diagram of the process flow for the production of biodiesel from waste oil.

The fresh methanol (METH stream) is mixed with recycled methanol (REC-MET stream) in a mixer (MIX2) to form the mixed methanol stream (M-METH stream), which consist mainly of methanol with trace amount of other recycled impurities from the recycle stream. Waste oil (WCO stream) with flow rate of 985 kg/hr is considered as a mixture of triolein and oleic acid (FFA). The compositions of FFA and the triglyceride (TG) is 50 wt% each. The waste oil (WCO stream) is mixed with the mixed methanol stream and fed to the reactor. The feed stream is pumped into the CSTR by a pump (P1) at high pressure. The solid catalyst is continuously fed into the CSTR with a rate of 4.9 kg /hr. It is equivalent to 0.005 mass ratio of the catalyst to waste oil. The methanol to waste oil molar ratio is fixed at 10.4 for waste oil with 50 wt% of FFA. The residence time in the CSTR is fixed at 2 hours since the reaction reaching its steady state at around 100 to 120 minutes [6].

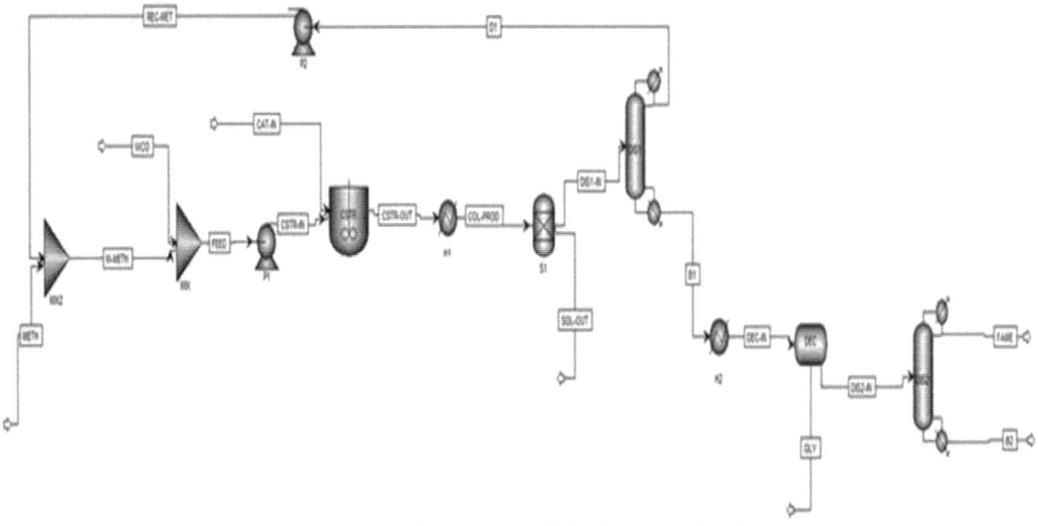

Figure 2. Heterogeneously catalyzed biodiesel production process.

3. Biodiesel process simulation

The process was simulated using Aspen Plus V8.6, as shown in Figure 2. All the components involved except for the catalyst are chosen from the components data banks available in Aspen. These components are water, methanol, glycerol, oleic acid (FFA), methyl-oleate (FAME), triolein, diolein, and monoolein. The catalyst, however, is input as a nonconventional solid in Aspen simulation as the molecular weight of the catalyst used (i.e. carbon-based solid acid catalyst from petroleum asphalt) is unknown. Since non-conventional solid is involved, the stream class for the simulation is chosen as MIXNCPSD. The base method for the simulation is chosen as NRTL for all unit operations as in West et al [3] except for decanter. NRTL is chosen because the components involved in the mixture are highly non-ideal due to the presence of high polar compound like methanol [3]. UNIF-DMD method is used for decanter in order to achieve the desired phase split [1,8]

4. Result and discussions

The details of all unit operation are shown in Table 1, while the details of the feeds and products stream are shown in Table 2. Table 3 shows the yields of both esterification and transesterification reactions. The yield of esterification is much higher than that of the transesterification. Shu et al [6] reported that triglyceride molecule with large alkyl chain is harder to activate compared to FFA. Their study on biodiesel production from high free fatty acid waste oil shows that esterification has higher activity compared to transesterification for all carbohydrate-based solid acid catalyst used in their study. The low yield of transesterification is due to the relatively low selectivity of FAME. On the other hand, the yield of FAME from esterification is higher. Therefore, large amount of FAME is produced, which drives the equilibrium of the transesterification reactions towards the reactant side, causing a low yield in the transesterification of oil [6]. In order to achieve higher yield, sensitivity analysis and optimization of the process towards several significant reaction variables is also carried out.

4.1. Sensitivity analysis

The flow rate of FFA in WCO stream is varied from 300 kg/h to 492 kg/h, which is equivalent to 38 wt% to 50wt% of FFA in the WCO feed. The decrease in FFA composition causes a slight drop in the yield of both transesterification and esterification (see Figure 3). The changes in both yields are about one percent for a 10% decrease in FFA composition. The increase in FFA composition drives the equilibrium of esterification towards the products side, which causes the slight increase in the esterification yield. The rate of transesterification is much slower compared to the rate of esterification due to higher activation energy and the complexity of transesterification. Therefore, high FFA is favorable for simultaneous esterification and transesterification. The same trend is reported in the study carried out by Li et al. [9]. FAME yield (Y_T) from transesterification is given by;

$$Y_T = \frac{\%\,conversion\,of\,TG}{100} \times \frac{\%\,selectivity\,for\,FAME}{100} \times 100\%$$

Table 1. Operation details of all unit operations

Process	Operating Details	
Reaction	Reactor type	CSTR
	Reaction Involve	Transesterification & Esterification
	Reaction Type	Second order kinetic
	Catalyst	Carbon based solid acid catalyst
	T, P	200 °C, 40 bar
	Methanol to oil ratio	10.4
	Residence Time (hr)	2
Methanol Recovery	Reflux ratio	1.1
	Number of Stages	14
	Feed Stage Position	5
	Distilate to Feed ratio	0.8
	Distillate purity (%)	99.71
	Top/Bottom Stage Temperature	64.3/126.7
Catalyst Separation	Separator Type	Solid-Liquid Separator
	Solid Split Fraction	1
Glycerol & Water Separation	Mechanisms	Gravity Separation
Biodiesel Purification	Reflux ratio	1.2
	Number of Stages	19
	Feed Stage Position	10
	Distillate to Feed ratio	0.79
	Distillate purity (%)	99.1

Table 2. Feed and product stream details

	Feed Stream				Product Stream			
	METH	WCO	CAT-IN		FAME	B2	GLY	SOL-OUT
Phase	Liquid	Liquid	Solid	Phase	Liquid	Liquid	Liquid	Solid
Temperature (°C)	25	25	25	Temperature (°C)	60.98	248.72	23.58	25
Pressure (bar)	1.013	1.013	1.013	Pressure (bar)	1.013	1.013	1.013	1.013
Molar Flow (kmol/hr)	2.96	2.30	-	Molar Flow (kmol/hr)	2.81	0.75	1.70	-
Mass Flow (kg/h)	94.78	985.09	4.90	Mass Flow (kg/h)	775.65	265.36	38.90	4.90

Composition (mass fraction)				Composition (mass fraction)				
Methanol	1.00	0	-	Water	0.0036	0.0000	0.7124	-
Triolein	0	0.5	-	Methanol	0.0025	0.0000	0.0490	-
Oleic Acid	0	0.5	-	Glycerol	0.0001	0.0000	0.2386	-
				Oleic acid	0.0000	0.0521	0.0000	-
				Methyl-oleate	0.9912	0.2755	0.0000	-
				Triolein	0.0025	0.1081	0.0000	-
				Mono-olein	0.0000	0.5638	0.0000	-
				Diolein	0.0000	0.0005	0.0000	-

Table 3. FAME yield of transesterification and esterification

Reaction	Parameter	
Transesterification	Conversion of Triolein (%)	93.78
	Selectivity of FAME (%)	73.16
	Yield (%)	68.61
Esterification	Conversion of FFA (%)	97.19
	Yield (%)	97.19

Table 4. Comparison of Optimization results against base case

Variables	Optimization Results	Base case
FAME stream flow rate (kmol/hr)	2.8132	2.8129
Manipulated:		
Reactor Temperature (°C)	223.1	200
Constraints		
FAME purity (wt%)	99.0	99.1
Methanol purity at the top of DIS1 (wt%)	99.7	99.7
Temperature of DIS1 (°C)	125.1	126.7
Temperature of DIS2 (°C)	249.6	248.7

From Figure 4, it is seen that the increase in temperature leads to a decrease in FAME yield from transesterification. From the kinetic of the reactions, higher temperature favors the formation of monoglycerides [6]. This also explains why the monoglyceride in the simulated process has a much higher composition compared to diglyceride and glycerol. On the other hand, an increasing trend in FAME yield is found with increasing temperature in esterification reactor.

4.2 Optimization.

Table 4 shows that the FAME product flow rate is above 2.81 kmol/h for all tested temperature. The biodiesel flow rate is maximized considering CSTR temperature (180°C to 250°C) as a decision variable and FAME purity (>98%), Methanol recovery purity (>98%), Temperature of DIS1 (<150°C) & Temperature of DIS2 (<250°C) as constraints. The optimized results show that a higher temperature of 223.1°C can produce the maximum FAME product (see Table 4). However, the increase of the FAME product flow rate is insignificant. The use of higher temperature is coupled with the increase of pressure and thus it will result in higher operating cost.

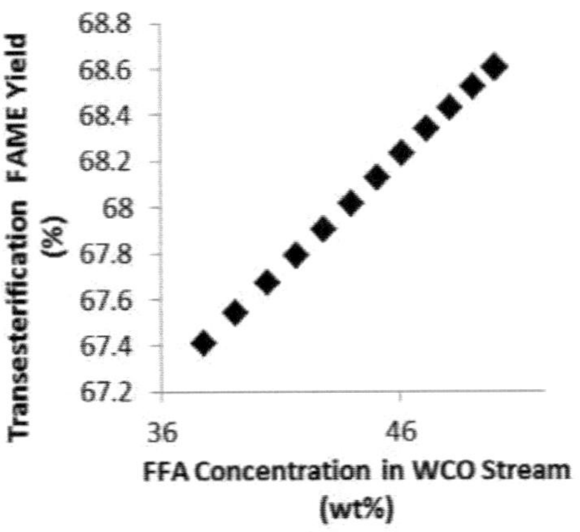

Figure 3a. Sensitivity Analysis by varying FFA concentration in WCO stream in transesterification

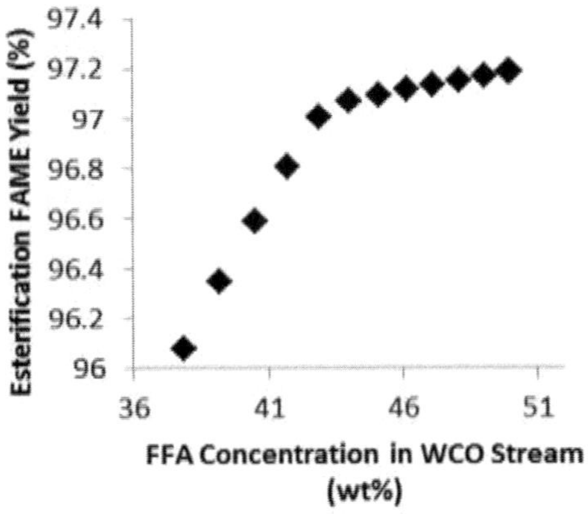

Figure 3b. Sensitivity analysis by varying FFA concentration in esterification.

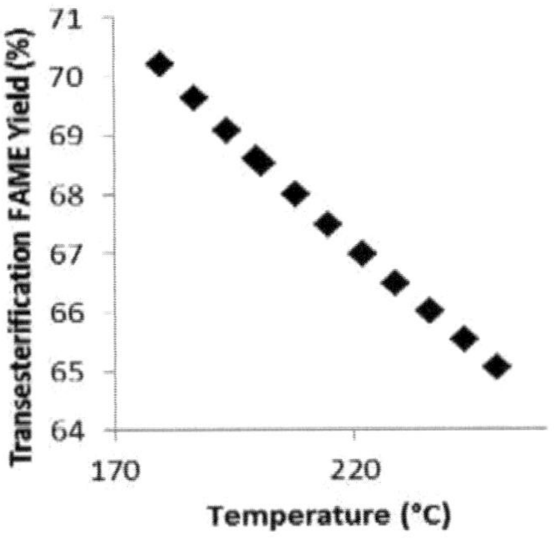

Figure 4a. Sensitivity Analysis by varying reactor temperature in transesterification

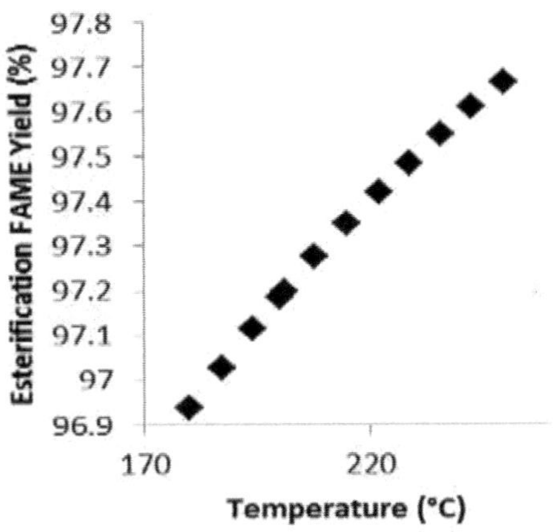

Figure 4b. Sensitivity analysis by varying reactor temperature in esterification.

5. Conclusions

A biodiesel production from the oil having 50% FFA using a heterogeneous catalyst is simulated using Aspen Plus V8.6. The process is relatively simple, and has lesser number of unit operations. Separation of catalyst can be done using a simple solid-liquid phase separator. The kinetics of simultaneous transesterification of triolein and esterification of FFA using a carbon based solid acid catalyst are considered in the simulation. FAME with purity of 99.12wt % at flow rate of 2.81 kmol/hr is produced from this process. The yield of FAME from transesterification and esterification is 68.61% and 97.19% respectively. Thus, the simulated process is feasible in producing FAME meeting the international standard (EN 14214). In addition, optimization and sensitivity analysis studies are carried out.

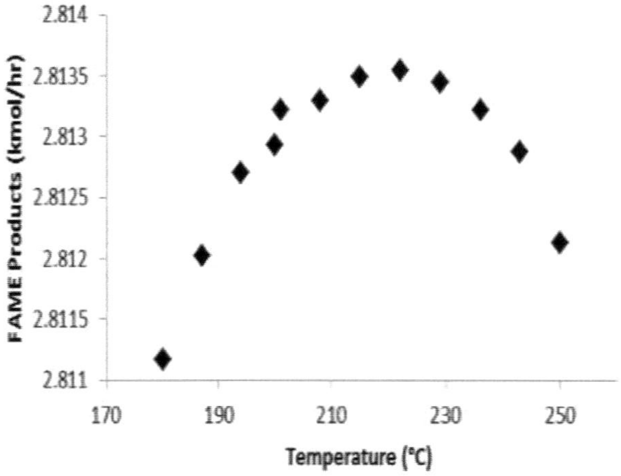

Figure 5. FAME products flow rate at different reactor temperature.

6. References

[1] Patle D S, Sharma S, Ahmad Z and Rangaiah G P 2014 Multi-objective optimization of two alkali catalysed processes for biodiesel from waste cooking oil *Energy Conversion and Management* **85** 361-372

[2] Phan A N and Phan T M 2008 Biodiesel production from waste cooking oils *Fuel* **87** 34903496

[3] West A H, Posarac D and Ellis N 2008 Assessment of four biodiesel production processes using HYSYS. Plant *Bioresource Technology* **99** 6587-6601

[4] Lam M K, Lee K T and Mohamed A R 2010 Homogeneous, heterogeneous and enzymatic catalysis for transesterification of high free fatty acid oil (waste cooking oil) to biodiesel: A review *Biotechnology Advances* **28** 500-518

[5] West A H, Posarac D and Ellis N 2007 Simulation, case studies and optimization of a biodiesel process with a solid acid catalyst *International Journal of Chemical Reactor Engineering* **5** Article A37

[6] Shu Q, Gao J, Liao Y and J. Wang 2011 Reaction kinetics of biodiesel synthesis from waste oil using a carbon-based solid acid catalyst *Chinese Journal of Chemical Engineering* **19** 163-168

[7] Serio M D, Mallardo S, Carotenuto G, Tesser R and Santacesaria E 2012 Mg/Al hydrotalcite catalyst for biodiesel production in continuous pasked bed reactors *Catalysis Today* **195** 54-58

[8] Patle D S, Ahmad Z and Rangaiah G P 2014 Plantwide control of biodiesel production from waste cooking oil using integrated framework of simulation and heuristics *Industrial & Engineering Chemistry Research* **53**(37) 14408-14418

[9] Li M, Zheng Y, Chen Y and Zhu X 2014 Biodiesel production from waste cooking oil using a heterogeneous catalyst from pyrolyzed rice husk *Bioresource Technology* **154** 345-348

Acknowledgments

The authors would like to acknowledge the financial support received from the Universiti Sains Malaysia through the Research University (RU) grant 1001/PJKIMIA/814155.

IOP Conf. Series: Materials Science and Engineering **121** (2016) 012008 doi:10.1088/1757-899X/121/1/012008

A study on variation in dissolved silica concentration in groundwater of hard rock aquifers in Southeast coast of India

K Pradeep[1]*, M Nepolian[1], P Anandhan[1], Chandran[1], R Kaviyarasan[1], M V Prasanna[2] and S Chidambaram[1]

[1]Department of Earth Sciences, Annamalai University, Annamalai Nagar, Tamil Nadu, India
[2]Department of Applied Geology, Faculty of Engineering and Science, Curtin University Sarawak, CDT 250, 98009 Miri, Sarawak, Malaysia

*Email: pradeep619rep@gmail.com

Abstract. Ground water of hard rock aquifers due to its lesser permeability results in the increased residence time, which leads to the higher concentration of ions. Hence in order to understand the hydro-geochemistry of the groundwater of a hard rock aquifer in India, 23 groundwater samples were collected from different locations of the study area and subjected to analysis of major cations and anions. The results of silica showed different range of concentration and was plotted in different groups. In order to understand the reason for this variation, different techniques like Thermodynamics, Statistics and GIS were adopted and it was inferred that the concentration was mainly governed by lithology and land use pattern of the study area.

1. Introduction

Silica in ground water can be found in two forms as dissolved silica and in particulate matter. Silica dissolved in natural waters is considered to be a good indicator of weathering and water circulation conditions [13]. Silica released as a result of chemical weathering of silicate minerals in rocks and sediments acquired by circulating groundwater and therefore the source of silica (SiO_2) in groundwater is mainly from water– rock interaction [15]. The amount of silica released into the water is conditioned by different factors, such as water saturation deficit of the aeration zone, seasonal fluctuations of precipitation and temperature, bedrock reactivity, and mineral stability [13]. Dissolved silica in water is normally measured as SiO_2 despite the fact that the exact form of dissolved silica is not known [1]. Average concentration of silica in natural water ranges between1.0 and 30.0 ml/L [12].

The solubility limit for silica in water is estimated at approximately 120 mg/L at 25°C [40][37]. Solubility of silica in water is directly proportional to temperature [16]. Fournier and Giggenbach et al. (1985) has studied and reported that temperature–solubility relationship of silica and they concluded that at least a temperature of about 300°C is used as a geothermometer in the studies of deep groundwater and geothermal systems. Another study [26] demonstrates that silica content of groundwater increases due to increased contact with silicate rocks and that the silica content is directly proportional to the residence time of water in host rock. They also showed that water ascending from deeper reservoirs had higher silica values than in groundwater of shallow origin.

The current research focuses mainly to understand the relationships between bedrock and groundwater chemistry. Groundwater chemistry has been investigated in Archean and Quaternary formations. The paper deals with various conditions that favor the silica concentration in groundwater, with a special emphasis on silica solubility in terms of the stability diagrams. Further, it also focuses on variation in the spatial distribution of silica concentration in groundwater of Archean and Quarternary formations.

IOP Conf. Series: Materials Science and Engineering **121** (2016) 012008 doi:10.1088/1757-899X/121/1/012008

2. Study area

The study area is located in TindivanamTaluk, of Villupuram district in figure 1. The study area is in the northern part of Tamil Nadu and close to the State capital of Chennai at a distance of about 100 Kms. Tindivanamtaluk is a hard rock region covering an area of 574.3 sq.km which lies between Lattitude 12° 22' - 11° 53'N and longitude 79° 29' - 80°1' E. It falls in survey of India Toposheet No. 58-M-6. The Tindivanam in general is having an average elevation of about 42.39 m above the mean sea level (MSL). The elevation map is show in the figure.2 higher elevation >85 m is in Northwestern part and lesser elevation of 13.89 m is in Southeastern region, Tindivanam had a population of 72,796 (Census, 2011) with a sex-ratio of 1,003 females for every 1,000 males, much above the national average of 929. Most of the study area is covered by agricultural lands which is main source of income for people in this taluk shown in figure.3. Numerous small surface water bodies can be seen and which are mainly helps to irrigation activities. It is also can be seen in figure.3 that salt pan activities in south western part of the study area. Salt exploration from back waters for domestic purpose is another source of income for the people at coastal area.

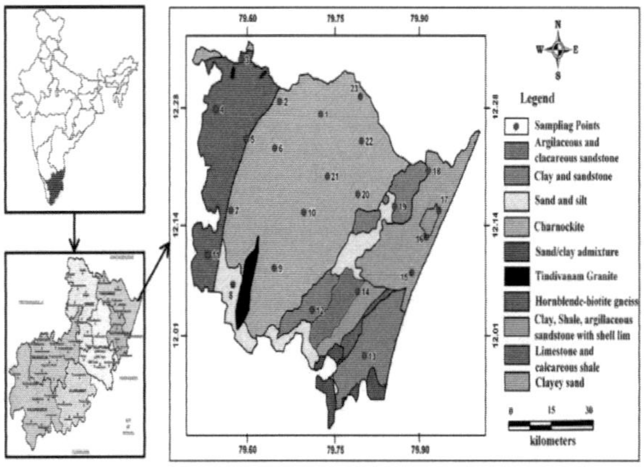

Figure 1. Land use and land cover map for the study area

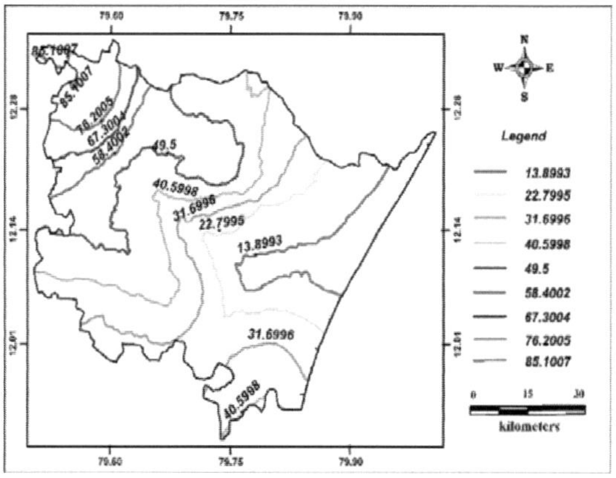

Figure 2. Surface elevation contours of the study area

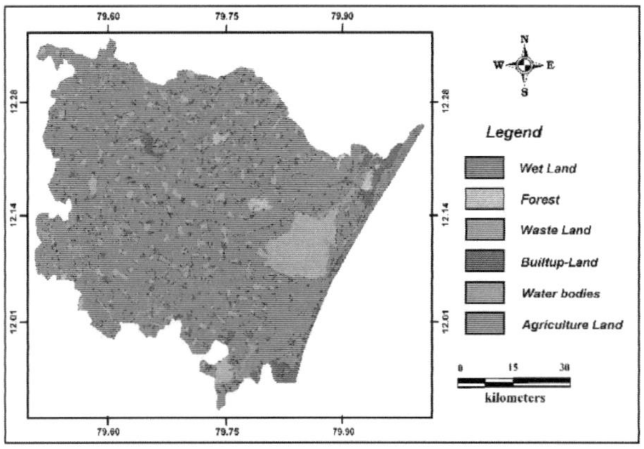

Figure 3. Land use and land cover map for the study area

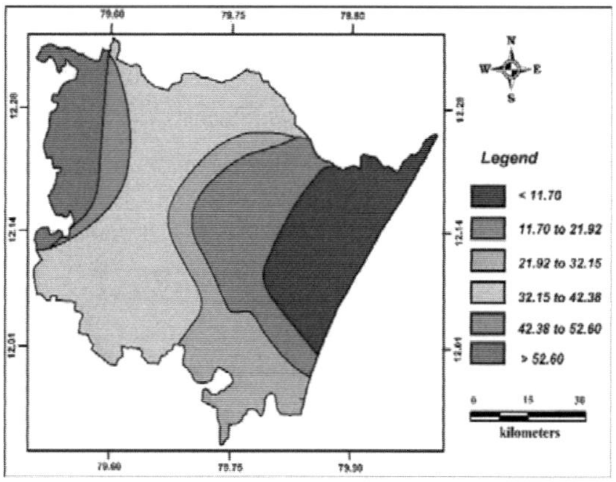

Figure 4. Spatial distribution of water level in the study area

2.1 *Climate and Rainfall*

The area falls under tropical climate with higher temperature in the summer months of March to May. The average temperature varies from 26 to 41° C [5]. Study area receives rainfall from southwest monsoon (June – September), and northeast monsoon (October – December). The higher rainfall is observed during the northeast monsoon period. The normal annual rainfall is 986 mm of decade. The average groundwater level varies from 1.47 m to 62.83 m above mean sea level (MSL) in the study area [5]figure 4.

2.2 *Geology*

Figure.1 shows the sampling points and spatial variation of geology in the study area. The study area predominantly comprises of Charnockite, and Hornblende-biotite gneiss in the western part of the study area. Pegmatites contain feldspar and smaller amount of quartz in some locations. Eastern part is mainly composed of quaternary formations. Northeastern part is covered by Clayey sand. Southern part is covered by Argillaceous and calcareous sandstones along with small amount of sand and silt deposits.

3. Methodology

A total of about 23 groundwater samples were collected as figure 1 from hand pumps representing entire TindivanamTaluk during the monsoon season. The parameters such as temperature and pH were measured in the field. Electrical conductivity (EC) and total dissolved solids (TDS) were measured in the laboratory by using standard electrodes. Collected groundwater samples were analyzed for its major anions and cations using standard procedures [3][34][35].

Calcium, magnesium, bicarbonate, and chloride were determined by trimetric method. Sodium and potassium were analyzed through flame photometry (ELICO CL 378). Silica, phosphate, and sulfate were determined by spectrophotometry (HACH Spectrophotometer). The reliability of the results were determined by the ionic balance of groundwater samples and a 5–10 % of percentage error was noted [14].

An aqueous model program WATEQ4F [4] was used to derive the Log pCO_2 and Ionic strength of groundwater samples. The statistical software SPSS-20 was used for correlation and factor analysis. Thermodynamic stability diagrams were plotted using WATCLAST program.

4. Results and Discussion

The analytical results along with maximum, minimum, and average values of chemical constituents are given in the Table1. Further the dominance orders of the ions are as follows:

$$Ca^{2+}>Na^{+}>Mg^{2+}>K^{+} = Cl^{-}>HCO_3^{-}>H_4SiO_4>NO_3^{-}>SO_4^{2-}>PO_4^{3-}$$

Groundwater occurs in the acidic to alkaline environment with a minimum pH value of 7.21 to a maximum value of 7.79. Higher electrical conductivity value noted along the course of the Gadilam River may be due to sewage infiltration [42]. Higher Ca^{2+} and Mg^{2+}occuring in groundwater may be due to weathering of primary mineral sources of rock water interaction [39]. Higher Na^{+} may be due to the weathering of plagioclase-bearing rocks [20]. K^{+} occurs in less quantity in groundwater which may be due to ion exchange or weathering process [10].

Piper diagram [30] was drawn by plotting the proportions (in equivalents) of the major cations (Ca^{2+}, Mg^{2+}, Na^{+}+ K^{+}) on one triangular diagram, the proportions of the major anions (CO_3^{-} + HCO_3^{-}, Cl^{-}, SO_4^{2-}) on another, and combining the information from the two lower triangular fields for cations, anions and a central diamond-shaped field shown in figure-5.

The groundwater can be classified into following category based on the location of points in diamond shaped field. In TindivanamTaluk, majority of the groundwater samples fall in Ca-Mg-Cl type and Ca-HCO_3 type, normal alkaline earth water with prevailing bicarbonate and chloride water type. The most of the samples fall in the Ca–Mg–HCO_3 type indicating the natural recharge and weathering from river or rain source or this facies may be due to rock – water interaction [33]. The most of the samples fall in the Ca–HCO_3facies indicating the dominance of freshwater recharge. This third type is predominantly due to Ion Exchanges [19] [23].

IOP Conf. Series: Materials Science and Engineering **121** (2016) 012008 doi:10.1088/1757-899X/121/1/012008

Table 1. Maximum, minimum and average of the chemical constituents in groundwater representing all samples (All values in mgL-1 except EC in μScm-1 and pH)

Parameter	*Maximum*	Minimum	*Average*
pH	7.79	7.21	7.54
EC	4010	421	1590.8
Ca	92	28	57.1
Mg	61.2	15.6	32.5
Na	72.2	2.9	27.1
K	10	0.3	3.3
Cl	135.3	31	85.31
HCO_3	274.5	81.5	167.3
SO_4	37.1	16.3	21.9
PO_4	86.548	6.422	15.489
H_4SiO_4	108.82	0.044	30.38
TDS	1850	220	720.5

4.1 Electrical Conductivity

The spatial distribution of EC gives three general trends in water as figure 6. Higher concentration of EC was noted in the southern part nearby Varahanadi River. This may be due to the infiltration of saline water from salt pan activities in this region. Lower concentration of EC was observed in the northern part. EC values for majority of the groundwater samples were varies between 1139 and 1857 μS/cm.

Most important among the weathering reactions is the incongruent dissolution of aluminium silicates which may schematically be represented by

$$\text{Cation + Al-silicate} + H_2CO_3 + H_2O = HCO_3^- + H_4SiO_4 + \text{cation + Al-silicate(s)} \quad (1)$$

Essentially, a primary mineral is converted into a secondary mineral. The secondary minerals are frequently structurally ill-defined or X-ray amorphous. The structural breakdown of aluminium silicates is accompanied by release of cations and usually of silicic acid. As a result of such reactions alkali in it is imparted to the dissolved phase from the bases of the minerals. In most silicate phases Al is conserved during the reaction, the solid residue being higher in Al than the original silicates. Because the alkalinity of the solution increases during the weathering processes, the solid residue has a higher acidity than the original aluminium silicate.

The stages of structural breakdown of minerals can be established by the stability field of the silicate minerals. Hence standard stability diagrams were evolved for the thermodynamic studies [17] and [40]. In spite of these limitations, a major application of these diagrams is that samples can be plotted to understand the appropriate ion activities for a given sample of water on these standard fields and evaluate the position of water composition in terms of mineral-water equilibria. Similar studies were adopted for composition of well waters in the black soil zone over weathered Deccan basalt of Malwa Plateau, Madhya Pradesh [25]. Studies were also reported to understand the ground water composition in lateritic soil profiles of Kerala is consistent with occurrence of Gibbsite, Kaolinite and halloysite [36] [8].

Minerals of the Kaolinite group are the main alteration products of weathering of feldspar. Besides Kaolinites and Montmorillonites, micas are possible products or intermediates. Mica (illite) has been identified as an intermediate in the decomposition of Na/K-feldspar.

Aggressive waters, high in CO_2 and low in dissolved solids encounter silicates high in cations and silica leaving Alumino- silicate residue with an increase in Al-Si ratio [5]. The initial water yields and attacks the Gibbsite residue which is evidenced by few points in Gibbsite field. As water continues to attack feldspar, pH rises with increase in cation and silica. Kaolinite forms until cations and silica contents rise high enough so that Montmorllionite initiates its formation.

Stability diagram have been widely used for understanding the geochemical behavior between minerals and water phase. In Na system, samples are stable with Kaolinite and it tends to move towards Montmorillonite due to the excess supply of silica figure 7. There are two distinct groups identified with lesser and greater silica falling in Kaolinite Montmorillonite field respectively and also indicating the movement of ions along the groundwater flow path figure 4. However, as shown by [38] maximum silica solubility at low temperature is controlled by amorphous silica rather than by quartz. The rate of crystallization of quartz is slow that amorphous silica, a metastable phase, should be looked upon as the upper limit of dissolved silica content of natural waters for most low-temperature processes.

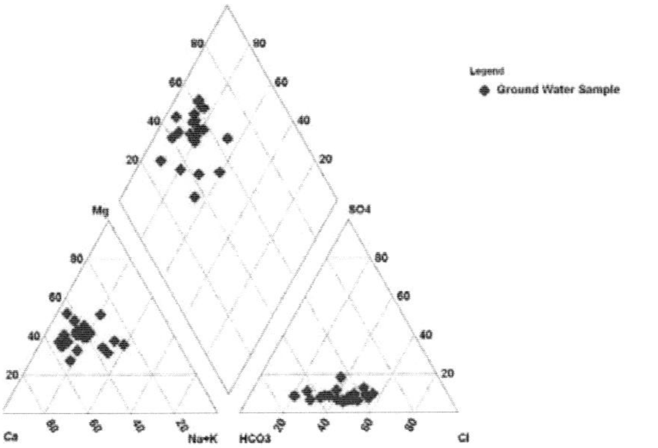

Figure 5. Piper diagram for groundwater samples from Tindivanam Taluk

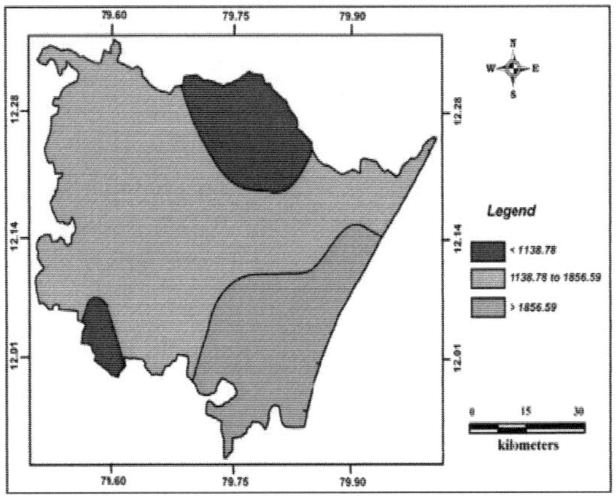

Figure 6. Spatial distribution of EC (µS/cm) in groundwater

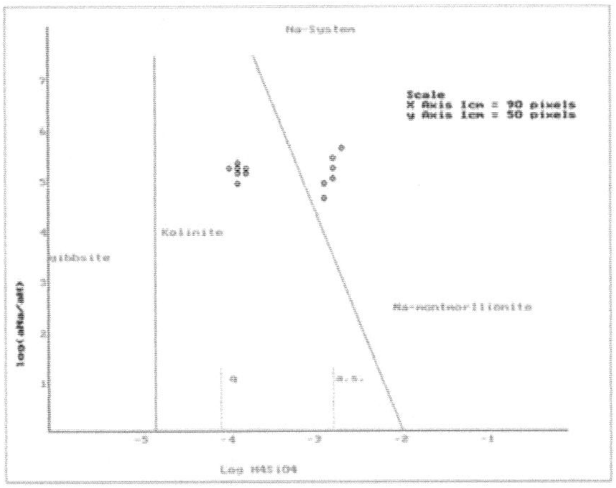

Figure 7. Thermodynamic stability plot of Na System

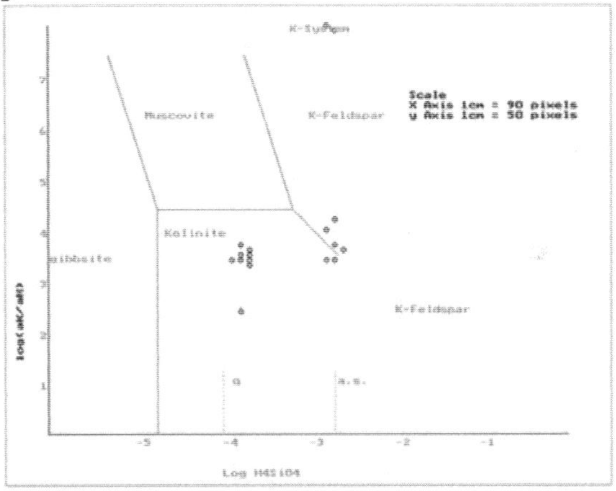

Figure 8. Thermodynamic stability plot of K System

In K system figure8 shows that samples are stable with Kaolinite and tend to move towards K-feldspar field, indicating the abundance of cations which may be associated with the mineral like Montmorillonite, Chlorite and Illite present in the aquifer system [33][8].

When reactions are written for the mineral compatibilities, it is discovered that only three variables need be considered: $[K^+]$, $[H^+]$, and $[H_4SiO_4]$. Furthermore, the ratio of $[K^+]$ to $[H^+]$ in the various equilibrium constants is always unity. Therefore, all the mineral relations can be described in two-dimensional representations involving the ratio of $[K^+]$ to $[H^+]$ as one axis, and the activity of H_4SiO_4 as the other.

The reactions and their equilibrium constants are as follows

$$3KAlSi_3O_8\,(c) + 2H^+\,(aq) + 12H_2O = KAl_3Si_3O_{10}\,(OH)_2\,(c) + 6H_4SiO_4(aq) + 2K^+(aq) \quad (2)$$

$$2KAlSi_3O_8c + 2H^+ (aq) + 9H_2O = H_4Al_2Si_2O_9 (c) + 2K^+aq + 4H_4SiO_4 (aq) \quad (3)$$

$$2KAl_3Si_3O_{10} (OH)_2(c) + 2H^+(aq) + 3H_2O = 3H_4Al_2Si_2O_9(c) + 2K^+(aq) \quad (4)$$

$$2KAl_3Si_3O_{10} (OH)_2(c) + 2H^+(aq) + 18H_2O = 3Al_2O_3.3H_2O(c)+ 2K^+(aq)+ 6H_4SiO_4(aq) \quad (5)$$

$$H_4Al_2Si_2O_9(c) + 5H_2O = Al_2O_3.3H_2O(c) + 2H_4SiO_4 (aq) \quad (6)$$

Among the aspects of interest is that most ground and stream waters fall into the Kaolinite field of stability. Also, they have dissolved silica contents ranging between the solubility of quartz and that of amorphous silica. The presence of silica in most waters in excess of that predicated for reactions with K-mica and K-feldspar can be attributed to the presence of other minerals that equilibrate more rapidly than K-feldspar.

In Ca system, samples tend to move from Kaolinite to Montmorillonite field figure 9. This indicates that the formation of new clay minerals in the area is mainly due to the supply of excess cations and silica to the pre-existing Kaolinite which appears to be formed owing to evaporation / evapo-transpiration process, as suggested by [22]. This may happen due to a long residence time of water in the aquifer matrix [39].

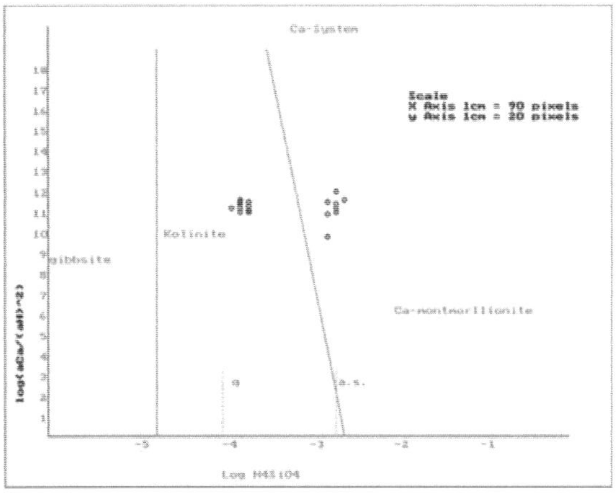

Figure 9. Thermodynamic stability plot Ca System

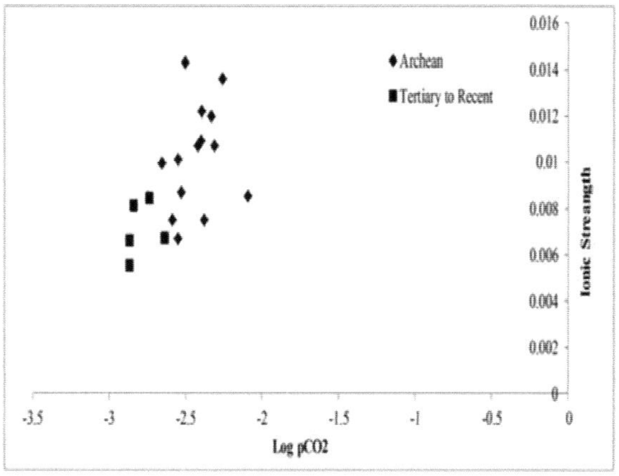

Figure 10. Relationship between log pCO_2 Vs. Ionic Strength

4.2 Partial pressure of Carbon Dioxide (CO_2) and Ionic Strength

Partial pressures of CO_2 (pCO_2) in water were calculated from measured pH and alkalinity of groundwater samples using the program WATEQ4F.

Log pCO2 were studied for two different formations exist in the study area. The log pCO_2 is ranges between -2.65 to -2.09 with an average of -2.43 in Archean formations. It ranges between -2.86 to -2.27 with an average of -2.60 in the Quaternary formations. It was observed that higher ranges of Log pCO_2 in hard rock aquifers.

The higher ionic strength indicates longer residence time or greater interaction with the host rock [9]. Most of the groundwater samples from Archean formations shown in figure 10 exhibits ionic strength greater than 0.01 and ranges between 0.01 and 0.0143 which shows the considerable interactions with host rock. The ionic strength of the Quaternary aquifer samples ranges from 0.005 to 0.01 which indicates that longer residence time or due to influence of salt pan activity and it is more in the topographic lows where water is stagnant due to surface elevations.

4.3 Dissolution of Silica minerals

4.3.1 pH and Silica

Hydrogen percentage plays an important role in rock weathering. The dissolution rates of most silicate minerals are independent of pH in the circumneutral region; they increase with decreasing pH in the acid region, and increase with increasing pH in the alkaline region [15]. It was observed that there is a correlation with pH and silica in certain groundwater samples from Archean formations figure.11. The alkaline nature of groundwater in Archean regions favors the dissolution of secondary minerals.

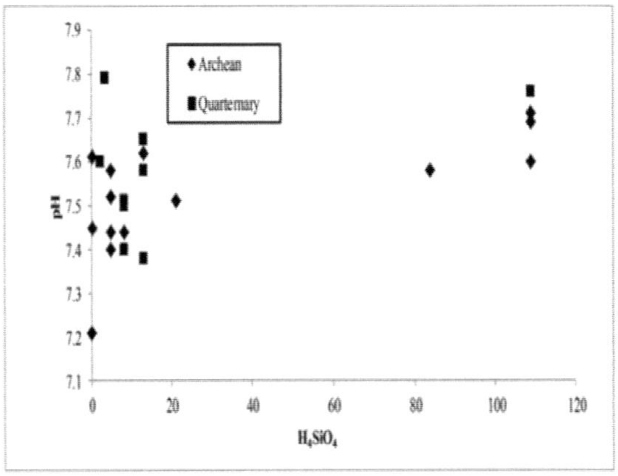

Figure 11. Relationship between H_4SiO_4Vs pH

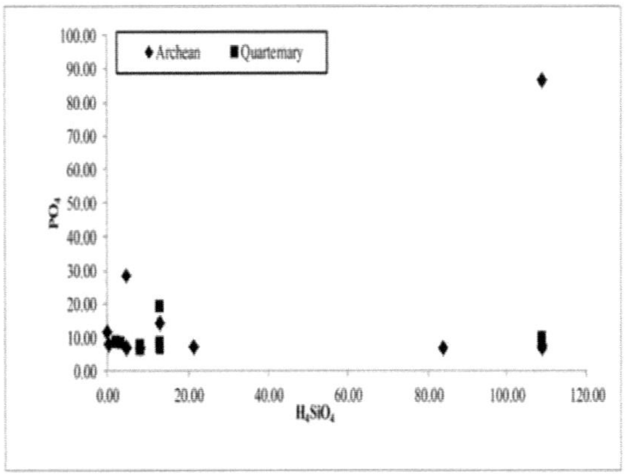

Figure 12. Relationship between H_4SiO_4Vs PO_4

4.3.2 Phosphate and Silica
In Archean formations the PO_4 concentrations are ranges between 6.75 and 86.55 with an average of 15.19 mg/L figure 12. The higher PO_4 concentration of 86.55 mg/L was observed in groundwater from Ilamangalam village. Majority of the samples from both formations show least amount of silica weathering whereas groundwater samples from topographically higher elevations in the western region show higher silica concentration.

4.3.3 Bicarbonate (HCO_3) and Silica
Weathering of silicate minerals leads the higher concentration of HCO_3 in groundwater [38] and it also indicates the process of recharge [28] [41] [42]. Figure13 shows that there is no definite relationship between HCO_3 and silica in groundwater from Archean formations. But in few samples HCO_3 concentration is increasing with increasing silica concentration. It can be noted from "Fig. 12 and 13" that concentration of PO_4 and HCO_3 for majority of the samples lies within the 20 mg/L of silica in groundwater.

4.3.4 Statistical Analysis
The need for statistical analysis of data for determining its hydrogeochemical nature is essential. Statistical data generally have better representation than graphical representations because (a) there are a finite number of variables that can be considered (b) variables are generally limited by convention to major ions and (c) superior relationships may be introduced by use of certain procedures [10].

4.3.4.1 Factor analysis
Factor analysis was carried out using SPSS-20 software package and the results reflect the complexity in chemistry. The factors were extracted with Eigen values >1 and their percentage of variance and cumulative percentage of variance of hydrogeochemical parameters of the groundwater are given in Table 2.
Factor 1 samples show strong positive loadings of Na^+, HCO_3, and Mg^{2+} with 23.03% of the total variance shown in Tables 2 and 3. The Na^+ is the dominant ion factor which is falls on the south eastern south western regions as figure 14. The higher Na^+ value is observed in the Idayanchavadi village. The representation of Na^+ in groundwater is may be due to leaching of secondary salts from backwater region [20] [8].
Factor 2 samples have strong positive loadings of Ca^{2+}, Mg^{2+}, and HCO_3^- with a total variance of 21.79 % Table 2 and 3. The dominant ions, Ca^{2+} and Mg^{2+}, observed in northwestern and south central part which may have originated from carbonate and silicate weathering along the contact between Charnockite and Hornblende-Biotite gneiss formations figure 14. Mg^{2+} in groundwater also liberated from the Mg-rich minerals like Enstatite–Hypersthenes, which is a more

common mineral in Charnockites / Gneissic formations [7] [10]. The representation of HCO_3^- is due to the weathering of silicates and it also indicates the process of recharge [39] [28] [41] [42].

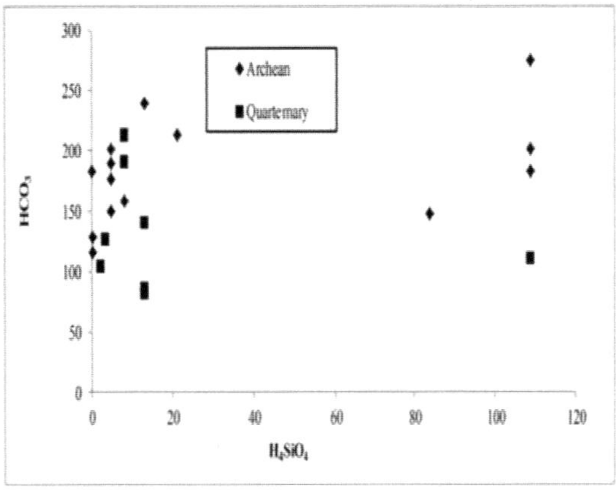

Figure 13. Relationship between H_4SiO_4Vs HCO_3

Table 2. Factor representing Eigen value >1, their percentage of variance and percentage of cumulative variance

Factor	Eigen values	Percentage of Variance	Percentage of Cumulative Variance
1	2.072	23.025	23.025
2	1.961	21.790	44.815
3	1.829	20.327	65.143
4	1.816	20.173	85.316

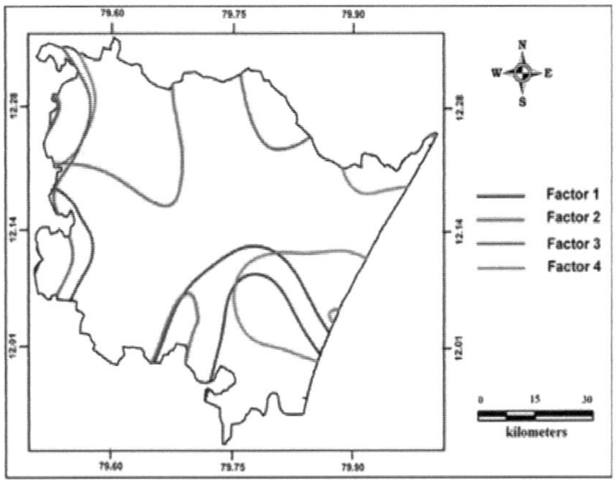

Figure 14. Contour map for factors observed in the study area

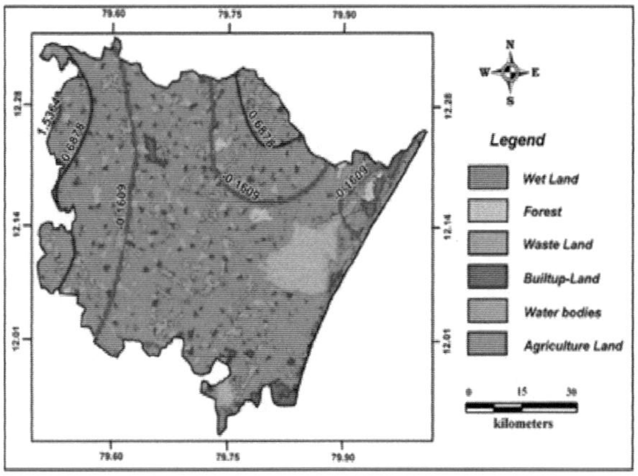

Figure 15. Comparison of Factor 3 with Land use and Land cover in the study area

Factor 3 samples have strong positive loadings of H_4SiO_4 with total variance of 20.33% Table 2 and 3. The western part representing more loadings of H_4SiO_4, this is mainly due to the dissolution of silicates where the land is mainly used for agricultural purpose figure 14. The northern part also shows small of amount of H_4SiO_4 loadings this also due to anthropogenic impacts from the agricultural practices [32][42]. It can be observed that the H_4SiO_4 is dominantly occurring in topographically elevated regions (western and northern parts) in the study area figure 2 and figure 15.

Factor 4 samples have strong positive loadings of K^+, SO_4^{2-} and PO_4^{3-} with a total variance of 20.17 %.The higher loadings of above factors were observed in the southern and eastern parts mainly due to agricultural impact on groundwater [41]. Sulfate and Phosphate in the groundwater system in this region is mainly due to use of considerable amount of Potash fertilizers [27] [11] [29] [2] [42].

The spatial representation of factors 1 to 4 shows that more number of factors (factor 1,2 and 4) were observed in south eastern part of the study area as figure 14 which are mainly influenced by leaching of secondary salts, weathering process and agricultural activities. Similarly, north western part also influenced three types of factors (factor 2, 3 and 4) predominantly by weathering process and agricultural activities. Groundwater flows mainly from western to eastern

IOP Conf. Series: Materials Science and Engineering **121** (2016) 012008 doi:10.1088/1757-899X/121/1/012008

direction figure 4. Water is tapped in the lower part due to salt pan activity and backwater regions which are greatly helps to leaching of secondary salts.

Table 3. Rotated component matrix

	1	2	3	4
Ca	-.067	.893	-.282	-.091
Mg	.589	.630	-.113	.286
Na	.900	-.075	-.109	-.088
K	-.218	.037	.041	.885
Cl	.454	.424	-.682	.042
HCO_3	.659	.599	.288	-.044
SO_4	.425	-.131	-.185	.805
PO_4	-.186	.441	.484	.530
H_4SiO_4	.086	-.102	.952	-.043

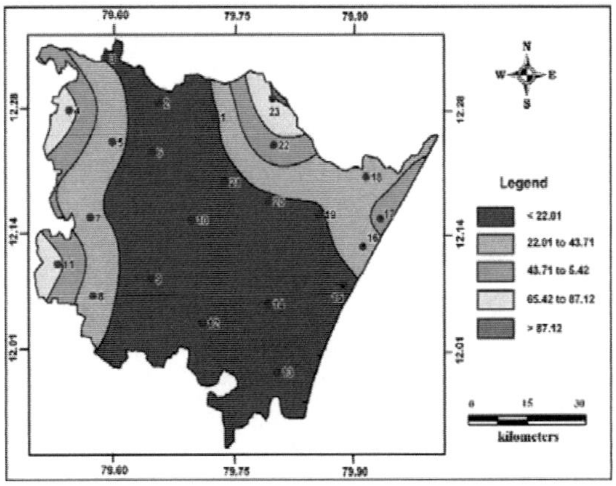

Figure 16. Spatial distribution of H_4SiO_4

It was observed from spatial distribution map for silica content in groundwater that higher silica in western and small patches of north eastern regions fig 16. It was observed from the study that three different factors (i.e natural and anthropogenic processes) are helping the increase the amount of silica in groundwater. It is mainly due to the weathering of the hard rock aquifer in the elevated region aided by the agricultural activity. The higher PO_4 and silica in figure 12 from high land in Archean formations show that there is dissolution of silica due to agricultural activities. Opaline silica from the rice plants provides a readily soluble source of silica and frequent oxidation and reduction cycles of paddy soil may accelerate weathering, thereby making silica more soluble [9].

5. Conclusion

Groundwater in TindivanamTaluk occurs in the near neutral to alkaline environment with a minimum pH value of 7.21 to a maximum value of 7.79. The most of the samples fall in the Ca–Mg–HCO_3 type indicating the natural recharge from surface water or this facies may be due to rock – water interaction. In Na system, samples are stable with Kaolinite and it tends to move towards Montmorillonite due to the excess supply of silica, and also indicating the movement of ions along the groundwater flow path. The alkaline nature of groundwater in Archean regions favors the dissolution of secondary minerals. The higher PO_4 and silica from high land in Archean formations show that there is dissolution of silica due to agricultural activities. The weathering enhances the HCO_3 concentration along with increasing silica concentration. The spatial representation of factors 1 to 4 shows that more number of factors (factor 1,2 and 4) are represented in south eastern part of the study area which are mainly influenced by leaching of secondary salts, weathering process and agricultural activities. Similarly, north western part is also influenced by three types of factors (factor 2, 3 and 4) like weathering process and agricultural activities. The dominant hydrogeochemical process in the study area is as weathering processes, leaching of secondary salts, and agricultural impact. The study concludes that three different factors (lithology, elevation and agricultural activity) are helping to increase the amount of silica in groundwater in Tindivanam Taluk.

Reference

[1] Al-Rehaili M and Abdullah., 2003 Comparative chemical clarification for silica removal from RO groundwater feed.*Desalination***159**. Pp.2 1-31.

[2] Amadi AN, Olasehinde PI, Yisa J, Okosun EA, Nwankwoala HO, Alkali YB 2012 Geostatistical assessment of groundwater quality from coastal aquifers of Eastern Niger Delta, Nigeria. *Geophys J Roy Astron Soc***2(3)**:51–59

[3] APHA 1998 Standard methods for the examination of water and wastewater, 19th edition. APHA, Washington DC, USASS

[4] Ball, J W and Nordstrom D K 1992 User's manual for WATEQ4F, with revised thermodynamic data base and test cases for calculating speciation of minor, trace and redox elements in natural waters. U.S. Geol. Surv., Open File Rep. 91-183, 189 pp

[5] Carrels R M and Mackenzie F T 1967 Origin of the chemical composition of some springs and lakes. In Equilibrium Concepts in Natural Water Systems; *Amer. Chem. Sot. Adv. Chem. Ser.* **67**, 222-242.

[6] CGWB 2009 West Bengal special issue. Quarterly journal of the central ground water Board. India: central ground water Board, ministry of Water Resources. Bhu-Jal News **24(1)**: 1-96. Available at http;//cgwb.gov.in/ documents/Bhujal_news_24_1.pdf (accessed in October, 2012).

[7] Chan HJ 2001 Effect of land use and urbanization on hydrochemistry and contamination of groundwater from Taejon area, Korea. *J Hydrol***253**:194–210

[8] Chidambaram S, Karmegam U, PrasannaMV, Sasidhar P, Vasanthavigar M 2011 A study on hydrochemical elucidation of coastal groundwater in and around Kalpakkam region, southern India. *Environ Earth Sci***64**(5):1419–1431

[9] Chidambaram S, Ramanathan AL, Prasanna MV, Anandhan P, Srinivasamoorthy K, Vasudevan S., (2007). Identification of hydrogeochemically active regimes in groundwaters of erode district, Tamilnadu-A statistical approach. *Asian J Water Environ Pollut***5**(3):93–102

[10] Chidambaram S2000 Hydrogeochemical studies of groundwater in Periyar district, Tamilnadu, India. Unpublished Ph.D thesis. Department of Geology, Annamalai University, Annamalai Nagar, Tamilnadu.

[11] Craig E, AndersonMP 1979 The effect of urbanization of groundwater quality. A case study of groundwater. Groundwater **17**(5):456–562

[12] Davis N S 1964 Silica in streams and groundwater. *Am. J. Sci.,***262**, 870–891.

[13] Dobrzynski Dariusz 2005 Silica origin and solubility in groundwater from the weathered zone of sedimentary rocks of the Intra-Sudetic Basin, SW Poland., *Acta Geologica Polonica*, *Vol. 55, No.4*, pp. 445-462.

[14] Domenico PA, Schwartz FW 1998 Physical and chemical hydrogeology, Wiley, New York,2nd edn, p 506

[15] Drever J I and Vance G F1994 Role of soil organic acids in mineral weathering processes.In The Role of Organic Acids in Geological Processes (ed. M. D. LEWAN and E. D. PITTMAN).*Springer-Verlag (inpress).*

[16] Fournier, R O 1983 A revised and expanded silica geothermometer. *Bull. Geotherm. Res. Council*, **11**, 329.

[17] Garrels RM, Christ CL 1965. Solutions Minerals and Equilibria. Harper and Row, New York, USA

[18] Giggenbach, WF, Gonfiantini, R, Jangi, B L and Truesdell A H 1985 Isotopic and chemical composition of Parbati valley geothermal discharges, Northwest Himalaya, India. *Geothermics*, **12**, 199–222

[19] Hardie LA, Eugster HP 1978 The evolution of closed-basin brines. *Min Soc Am Spec Publ***3**:273–290

[20] Hem JD 1985 Study and interpretation of the chemical characteristics of natural water. *U S Geol Surv Water Supply Pap***2254**:117–120

[21] Hem, J D 1986 Study and interpretations of chemical characteristics of natural water. *USGS Water Supply Paper*, **2254**, p. 263.

[22] Jacks G 1973 Chemistry of groundwat er in a district in Southern India. *J Hydrol***18**:185–200

[23] Jankowski J, Jacobson G 1989 Hydrochemical evolution of regional groundwaters to playa brines in Central Australia. *J Hydrol***108**:123–173

[24] Karthikeyan A and Senthil Kumar G R2011. Groundwater level fluctuation studies in part of tindivanam taluk, villupuram district, tamilnadu *international journal of recent scientific research*, vol. 2, issue, **11**, pp. 271 -275.

[25] Lunkad, S K and Raymahashay BC 1978.groundwater quality in weathered Deccan basalt of Malwa plateau, India. *Quart. J. Engineering Geol.*, U.K., **11**, pp. 273-277.

[26] Marchand D, Rayan M C, Bethune D N and Chu A 2002 Groundwater–Surface Water Interaction and Nitrate Origin in Municipal Water Supply Aquifers, Sanjose, Costa Rica.

[27] Miller GT 1979 *Living in the environment.*Wadsworth, Belmond, CA, 470p

[28] Ophori DU and Toth J 1989 Patterns of groundwater chemistry, Ross Creek Basin, Canada. *Ground Water***27**:20–26

[29] Papatheodorou G, Demopoulou G and Lambrakis N 2006 A longterm study of temporal hydrochemical data in a shallow lake using multivariate statistical techniques. *Ecol Model***193**:759– 776

[30] Piper, A M. A graphic procedure I the geo-chemical interpretation of water analysis, USGS Groundwater Note no, 1953, 12.

[31] Plummer LN, Jones B F and Truesdell AH 1976 WATEQF-A FORTRAN IV version of Wateq, a computer program for calculating chemical equilibrium of natural water, U.S. Geo. Sure. *Water-Resour. Invest.*, 76-13. Pp. 63.

[32] Prasanna MV, Chidambaram S, Shahul Hameed A, Srinivasamoorthy K 2009 Study of evaluation of groundwater in Gadilam basin using hydrogeochemical and isotope data. *Environ Monit Assess* **168**:63–90

[33] Prasanna MV, Chidambaram S, Srinivasamoorthy K 2010 Statistical analysis of the hydrogeochemical evolution of groundwater in hard and sedimentary aquifers system of Gadilam river basin, South India. *J King Saud Univ Sci* **22**(3):133–145

[34] Ramanathan, A.L, 1992 Geochemical studies in the Cauvery river basin. Unpublished Ph.D. Thesis, Punjab University Chandigarh.

[35] Ramesh R and Anbu M 1996 Chemical methods for environmental analysis—water and sediments. *Macmillan Publisher*, India, p 161

[36] Raymahashay B C, Rao KS, Mehta, VK. and Bhavana PR 1987 Mineralogy and geochemistry of lateritic soil profiles in kerela, India, *Chem. Geol.,* **60**, pp 327-330.

[37] Sanks R.I., Water treatment plant design, Ann Arbor Science, 1978.

[38] Siever, Raymond, 1957. The silica budget in the sedimentary cycle : *Am. Mineralogist*, V. **42**, pp. 821-841.

[39] Srinivasamoorthy K, Chidambaram S, Prasanna MV, Vasanthavigar M, John Peter A and Anandhan P 2008 Identification of major sources controlling groundwater chemistry from a hard rock terrain—a case study from Metturtaluk, Salem district, Tamilnadu India. *J Earth Syst Sci* **117**(1):49–58

[40] Stumm W and Morgan J J 1970 Aquatic Chemistry. Interscience Publishers, New York.

[41] SubbaRao N 2007 Groundwater quality as a factor for identification of recharge zones. *Environ Geosci* **14**:79–90

[42] Thivya C, Chidambaram S, Singaraja C, Thilagavathi R, Prasanna MV, Jainab I 2013 A study on the significance of lithology in groundwater quality of Madurai district, Tamil Nadu (India). Environ Development Sus. doi:10.1007/s10668-013-9439-z

[43] Stumm W and Morgan J J 1981 Aquatic Chemistry, John Wiley CyrS. ons, **2nd ed.**. 1981.

Acknowledgment
The authors acknowledge Mr.Chandran and Mr.Baskar for joint effort in collection of various water samples in the field and analyzing part in the laboratory. The authors also thank to Curtin University, Sarawak Malaysia for their support in the publication fee through Research Performance Index (RPI) fund of M.V Prasanna.

Assessment of sources for higher Uranium concentration in ground waters of the Central Tamilnadu, India

V S Adithya[1*], S Chidambaram[1], K Tirumalesh[2], C Thivya[1], R Thilagavathi[1] and M V Prasanna[3]

[1]Department of Earth Sciences, Annamalai University,Tamilnadu, India

[2]Isotope Production and Applications Division, Bhabha Atomic Research Centre, Mumbai, India

[3]Department of Applied Geology, Faculty of Engineering and Science, Curtin University Sarawak, CDT 250, 98009 Miri, Sarawak, Malaysia.

*E-mail: lechuadhi@gmail.com

Abstract. The uranium concentration in groundwater has attained greater importance considering the health effects in mankind. Groundwater being the major source of uranium; sampling and analysis of groundwater for the major cations and anions along with uranium has been carried out in hard rock aquifers of Madurai district. The sampling has been carried out in varied aquifers like, Charnockites, Hornblende Biotite Gneiss, Granites, Quartzites, Laterites and sandstone. The cation and anions showed the following order of dominance $Na^{+}>Ca^{2+}>Mg^{2+}>K^{+}$ and that of anions are $HCO_3^{-}>Cl^{-}>SO_4^{2-}> NO_3^{-}>PO_4^{3-}$. Higher concentration of uranium was found along the granitic aquifers and it varied along the groundwater table condition. Further it was identified that the mineral weathering was the predominant source of U in groundwater. Tritium studies also reveal the fact that the younger waters are more enriched in uranium than the older groundwater with longer residence time.

1.Introduction

The element uranium (U) is distributed throughout the crust of the earth in trace quantities in all the rock types. It is rich in acid igneous rocks like granites, syenites etc. while depleted in basic and ultra-basic rocks. The average concentration of U in Earth crust is 2.7 ppm [1]. In nature, U generally occurs in tetravalent state as insoluble species and hexavalent state as highly soluble species. The U concentration in groundwater depends on lithology, geomorphology and other geological conditions of the region [2 – 4]. Chemically, metallic uranium is highly reactive andcan be leached in acidic or alkaline environment. Uranium concentrations in most of the groundwaters are generally low, typically in the range of 0.1 to1 ppb, but it can

IOP Conf. Series: Materials Science and Engineering **121** (2016) 012009 doi:10.1088/1757-899X/121/1/012009

reach several tens to hundreds of ppb when it reacts with U rich minerals in the aquifers. Uranium is easily dissolved and transported in oxidizing groundwater due to the presence of oxygen; hence, it can be transported far away from its original occurrence [5]. Clay minerals and amorphous Fe, Tiand Mnoxides and hydroxides adsorb U from late-stage and post magmatic fluids, or are precipitated during supergenealteration and weathering of the primary minerals [6].

There are seven or more factors, including source rock U content, which can influence the uranium dissolved in water. These are

(1) The uranium content in source rocks, sediments or soils and its leachability.

(2) The proximity of the water to uranium-bearing rocks or minerals.

(3) The degree of hydraulic isolation of the water from dilution by fresher surface or subsurface waters.

(4) Climatic effects and their seasonal variability, particularly the influence of evapotranspiration.

(5) The pH and oxidation state of the water.

(6) Concentrations of carbonate, phosphate, vanadate, fluoride, sulfate, silicate, calcium, potassium and other species which can form uranium complexes or insoluble uranium minerals.

(7) The presence of highly sorptive materials such as organic matter, ferric, manganese, and titanium oxy-hydroxides and clays.

These factors have either roughly constant effect.Fairly high uranium concentrations among vegetable foodstuffs were found in cauliflower,carrots,cucumber, whitechampignon and red peppers. Leaf-rich vegetables, herbs, sauerkraut, dill and parsley.The uraniumcontentsofwildmushroomsvarywiththe site, itsgeologicalorigin,andwithspecies[7]. Animal products contain lesser concentration of U. Uranium causes both chemical and radiological toxicity which is said to be affecting mainly organs like Kidney and lungs [8].

The proposed study area is a complex terrain with varied lithology from hard rock to sedimentary formation along with granitic intrusions. The study aims to understand the sources and distribution of total uranium in groundwater in the aquifers of this region.

2. Study area

Study area is located in Central part of Tamil Nadu comprising of parts of Madurai, Dindigul, Trichy, Pudukottai and Sivaganga districts. Study area occupies a total of 4311 square kilometers and falls between E 09.89-10.35 latitudes and N 78.03- 78.81 longitudes as shown in “figure 1”.

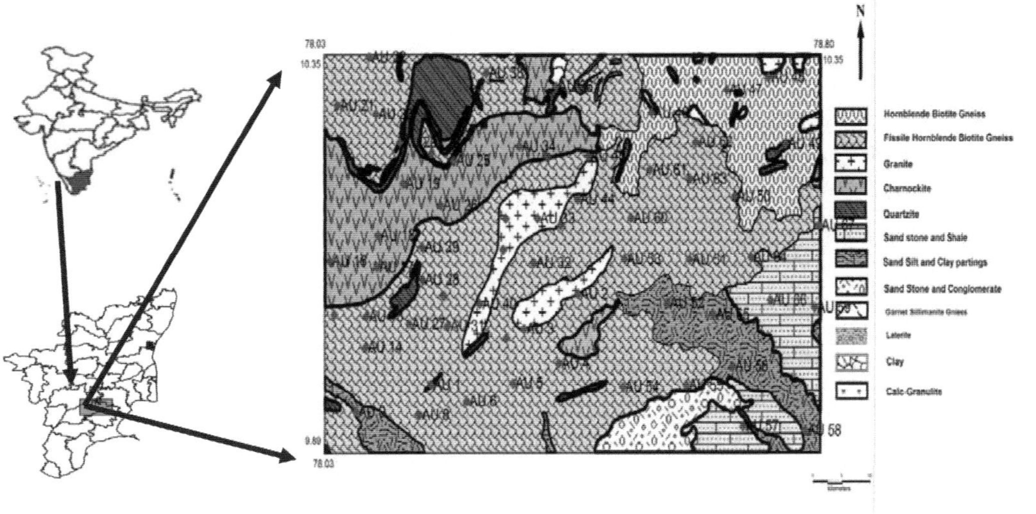

Figure 1.Location of study area showing lithology and sample locations.

IOP Conf. Series: Materials Science and Engineering **121** (2016) 012009 doi:10.1088/1757-899X/121/1/012009

Vaigai is a major seasonal river. Water level trend in 10 years from 1998 to 2007 showed fluctuation of 0.004 to 1.523 m/year in the study area by comparing the [9] CGWB (2008) reports of various districts, comprising of parts of Madurai, Dindigul, Trichy, Pudukottai and Sivagangadistricts.

3. Methodology

A total of 61 samples were collected spatially from hand pumps and bore wells of study area based on spatial coverage and lithology during the month of February 2014. The pH, temperature, total dissolved solids (TDS) and electrical conductivity (EC) of the water samples were measured in the field using a portable water-analysis kit. Sampling and analysis were carried out using standard procedures. Major ions such as Ca^{2+}, Mg^{2+}, Na^{+}, K^{+},F^{-}, Cl^{-}, CO_3^{-}, HCO_3^{-},NO_3^{-}, H_4SiO_4, $PO_4^{-}$and SO_4^{2-} were analyzed in lab. Cations such as Ca^{2+}, Mg^{2+} as well as anions such as Cl^{-}, CO_3^{-}, and HCO_3^{-} were analysed by titrimetry method. Na^{+} and K^{+} by flame photometer (ELICO CL 378).SO_4^{2-}, PO_4^{-}, NO_3^{-} and H_4SiO_4 were analysed by spectro-photometer (SL 171 minispec). EC, pH and F^{-} were measured using ion electrode meter (Thermo Orion). Precision of analysis was checked by error percentage calculation (1)

$$Error\ \% = \frac{(TA-TC)}{(TA+TC)} x100 \qquad (1)$$

(Where TA- total anions and TZ- total cations)

The percentage of error is found to vary between ±1 to ±10 %[10]. Total Uranium concentration in groundwater samples were analysed by Laser fluorimetry.

4. Results and Discussion

The total cations (TZ+) and total anion (TA-) balance [11] shows the charge balance error percentage. The error percentage is observed in the samples between ±10%. Occurrences of errors in chemical analysis of groundwater may also be due to reagents employed, limitations of the methods and instruments used, presence of impurities in distilled water etc. The maximum, minimum and average values for different seasons are given in table.1. The groundwater in the study area is generally odorless and colorless in most of the sampling locations.

IOP Conf. Series: Materials Science and Engineering **121** (2016) 012009 doi:10.1088/1757-899X/121/1/012009

Table 1. Minimum, maximum and mean of physico chemical parameters and comparison of chemical composition of water with WHO (2011) in milligram per litre. (ExceptU in ppb, ORP in mV, EC in µS/cm and pH)

Parameters	Maximum	Minimum	Average (N=61)	WHO (2011)
Ca	104	6	35.62295	100*
Mg	57.6	8.4	26.38689	50*
Na	158.9	2.6	44.75262	NG
K	40	0.1	8.888525	20*
Cl	116	2	39.90484	250
F	2	0.1	0.595082	1.5
HCO_3	286.7	24.4	156.3475	125-350*
NO_3	420	10	88.62295	50
SO_4	76	0.1	23.39013	500
H_4SiO_4	16.09	5.591	11.38392	NG
U	68.28	0.4	11.78	30
ORP (mV)	283	-270	48.54098	NG
pH	8.111	6.176	7.135541	6.5-9.5
EC(µS/cm)	4000	288.8	1420.632	1400*
TDS (mg/l)	1890	204	725.0984	1000

[a]NG- not given; * WHO [12]

Quality of drinking water was evaluated according to the WHO standards [8]. pH of majority of groundwater samples collected fall within safe limit, but the TDS of certain samples are higher than the maximum permissible limit. The lithological composition, leaching and dissolution of soil salts and chemical spills from wastewaters play an important role in the TDS values [13].More likely, many processes influence the relationship between pH and other physico-chemical parameters that even multivariate approach cannot identify them [14]. The large variation in EC is mainly attributed to lithologic composition and anthropogenic activities like disposal of sewage along the riverbeds prevailing in this region. The extremely low values of EC are attributable to short residence time of the groundwater in the rock matrix as suggested by [15] and [16] or due to the effect of dilution [17].

U concentration in groundwater is important in understanding the radiological impact to secure the standard of life. A number of studies have been carried out in different places in order to study uranium concentrations in drinking waters [18 -20]. The [8] standard limit for Uranium in groundwater is 30 ppb. In the study area 8 samples are above the permissible limit. Kidneys and bones are the major locates for the accumulation of uranium in the body[21- 23] and skeleton is the major site for the long term storage of Uranium [24]. Even lower concentration of Uranium in drinking water also causes the nephrotoxic effects [25].
The orders of dominance of U in different lithologies in various seasons are as follows
Granite > Fissile hornblende biotite gneiss >Charnockite>Hornblende BiotiteGneiss

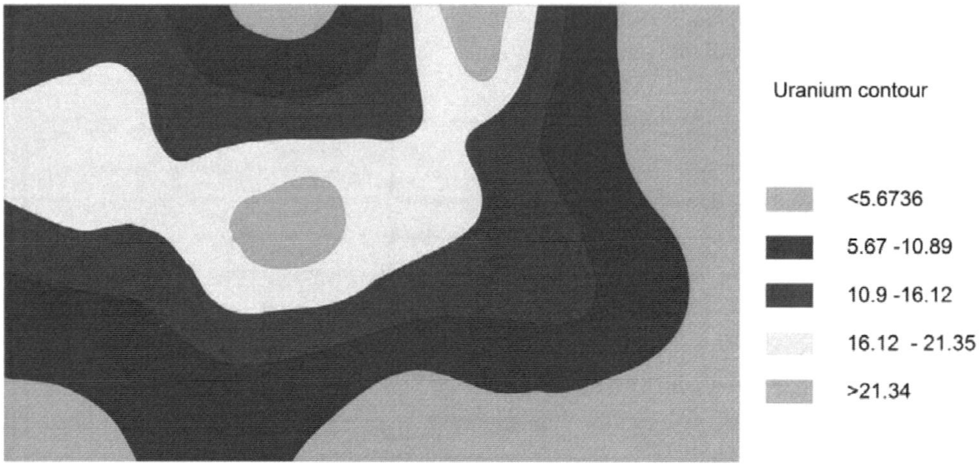

Figure 2. Spatial distribution of Uranium in groundwater samples of study area

Uranium and radon in ground water mainly depend on the fractures and their degree and the angle of inclination to the bedrock [26]. The high concentration of U in Central and north eastern part of the study area (Figure 2)could be due to the leaching of elements present in fractures. Temple (2002) [27] hypothesized that high U concentrations were due to U-rich fractures in the granitic rocks. Only relatively deep wells contained high U concentrations, possibly because U had already been leached from shallow fractures as it behaves as a closed system. The study area also reports presence of pink granitic intrusions [28] and the zone of higher U concentration is found around the graniticterrains that enhance the leaching of U to the groundwater system [29].

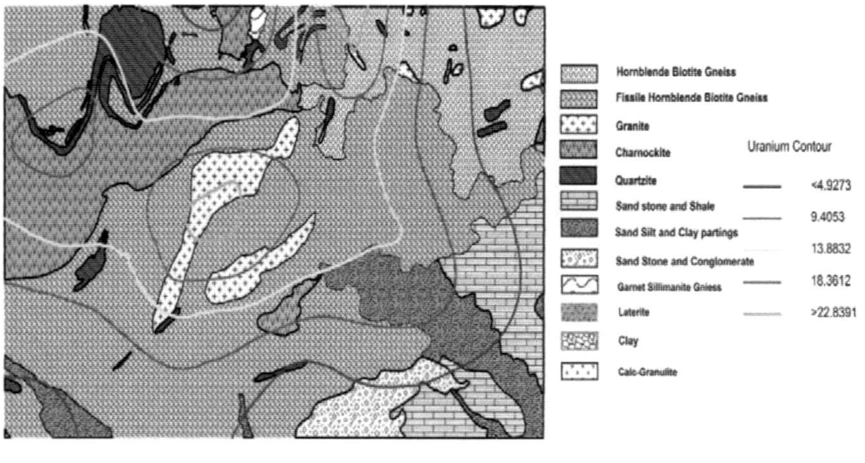

Figure 3. Spatial distribution of U overlaid on lithology map

High U concentration is generally reported in groundwater of granitic terrains(figure 3). A large range of U concentrations was observed in groundwaters from the Lac du Bonnet Granite of the Canadian Shield. U concentrations were found to be generally highest at shallow depths (<60 m; depth range of boreholes up to 1000 m) in oxic environment groundwaters which shows evidence of active groundwater circulation. The dominant controls on U concentration were redox potential and the formation of soluble anionic U-carbonate species. U increases with the accumulation of alteration phases along fractures and around the primary phases. As weathering progresses, this U is incorporated into the Interstitial spaces of biotite, feldspar and early alteration phases accumulate within the secondary interstitial aggregates and lose their discrete character [30].High-U groundwaters appear not to be a widespread feature of the granites of the region, perhaps because of the short residence times of the groundwaters and their slightly acidic, low-alkalinity compositions. The sporadic nature of the U mineralization may also be a factor.

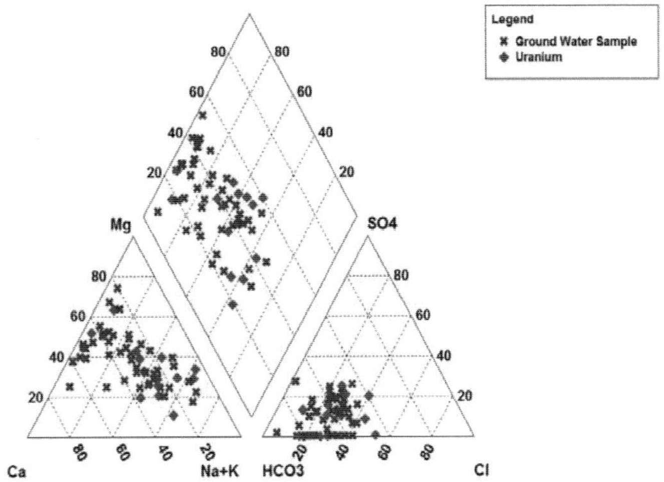

Figure 4. Piper facies for ground water samples highlighting samples with high uranium concentration

The higher value of U concentration which is above 30ppb is observed in few samples and their hydrogeochemical parameters is plotted in piper diagram (figure 4) to study the migration of U ions in groundwater samples. The samples are predominantly of HCO_3 type as weathering processes affect the mineral which in turn affects the distribution of elements in groundwater. Firstly Ca would have been removed from the groundwater, while Na which is absorbed on the surface of the clay minerals in the aquifer would be released into the groundwater. The Na probably comes from the decomposition of Plagioclase feldspar. An increase in Na and K is assumed to result from cation exchange of Ca and Mg ions by mineral-water interaction.

IOP Conf. Series: Materials Science and Engineering **121** (2016) 012009 doi:10.1088/1757-899X/121/1/012009

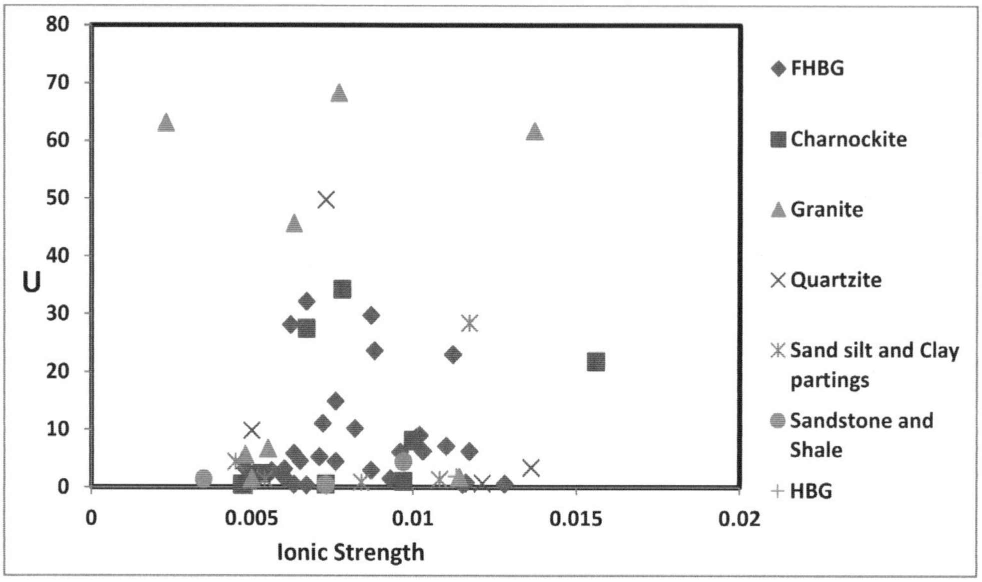

Figure 5. Scatter plot between total U and Ionic strength of the samples

The relationship between U and ionic strength (IS) is plotted in figure5.It is interesting to note that U concentration increases with increase of IS that may be attributed to the increase in residence time of water [31]. The higher U concentration in the groundwater samples shows that there is a decreasing trend with increase in IS. The study area being a hard rock terrain, residence time of water is generally higher in cracks and fissures if they are not interconnected sufficiently, which ultimately increases U concentration. Increase of ionic strength may also be due to pH induced weathering in the aquifer [32],as U concentrations in groundwater decreased with flow direction due to U-retention by sorption processes.

4.1. Factor Analysis

The purpose of Factor analysis is to simplify the quantitative description of a system by determining the minimum number of new variables necessary to reproduce various attributes of the data. These procedures reduces the original data matrix from one having (n) variables necessary to describe the (N) samples to a matrix with (m) factors (m < n) for each of the (N) samples. It is also aimed at transforming the variables so that the axes become orthogonal, which then allows the definition of new independent variables.The first factor is chosen to explain as much as possible of the total variance of the observations, the second factor to explain as much as possible of the residual variance, and so forth. In other words, the first factor is determined such that the sum of squares of the projections of the points on the factor is highest (factor loadings). Next, to define the second factor, the points are projected on a plane orthogonal to the first factor and so on for the other factors, each explaining less and less of the total variance. On the other hand, the sum of squares of the factor loadings for each variable is the communality and reflects the proportion of the total variability of each variable accounted for by the factoring. It follows three main steps, namely: extraction of initial factors, rotation of factors and calculation of each factor scores.

There were five different factors extracted for the analytical results (table 2) and the fifth factor represents U in ground water showing strong negative correlation with OPR, indicating that oxidation and reduction potential has a significant role in U concentration of ground waters apart from the weathering process.

IOP Conf. Series: Materials Science and Engineering**121** (2016) 012009 doi:10.1088/1757-899X/121/1/012009

Table 2. Rotated component matrix of groundwater samples.

	Component				
	1	2	3	4	5
Ca	.168	.800	-.016	-.162	.042
Mg	.400	.531	.481	.057	-.053
Na	.824	-.173	.149	-.006	.131
K	.786	-.199	-.095	-.186	.008
F	-.251	.223	.034	.775	-.145
Cl	.731	.345	-.210	-.075	.153
HCO_3	.866	.164	.043	.177	-.081
NO_3	.139	.223	.896	-.054	.040
SO	.465	.192	-.756	.014	.044
H_4SiO_4	-.241	.593	.091	.130	-.079
U	.171	-.087	.076	.243	.784
ORP	.088	-.047	.090	.378	-.749
pH	.209	-.337	-.123	.661	.140

U must be oxidized before it is transported into groundwater but once in solution, it can travel great distances [33], hence it can be transported far away from its original emplacement [5]. In deeper environments, mobility and attenuation of U are controlled by the composition of fracture coatings and water chemistry. In general, the highest concentration of dissolved U is found in the oxidized zone of the aquifer. As groundwater conditions become reduced, dissolved U concentration decreases due to sorption and precipitation reactions. Deep aquifer systems are recognizable by a constant low U concentration(figure 6).

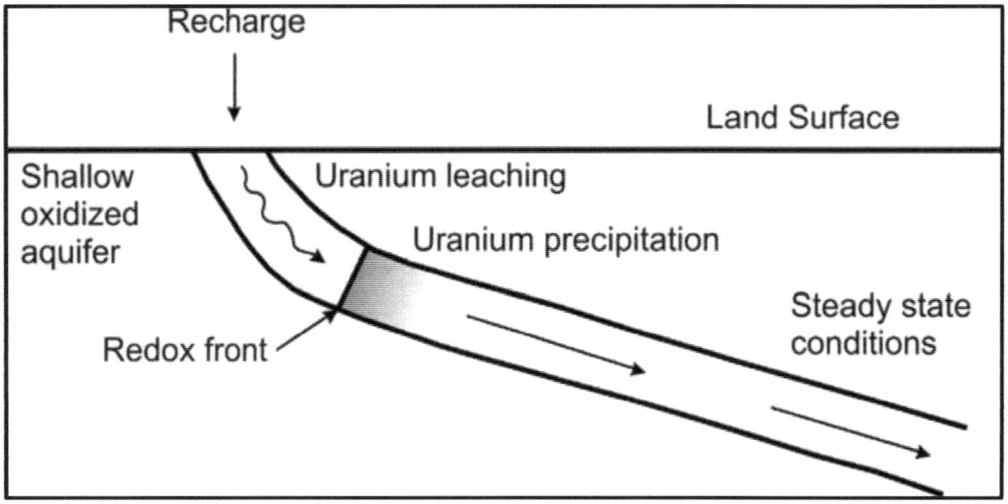

Figure 6. Schematic diagram of U concentration distribution along a groundwater flow path(Revised fromIvanovich et al)[34]

4.2. Factor Score

Factor scores were estimated to find out the spatial representation of the factor and its temporal variation [35]. They are commonly obtained by two approaches such as weighted least square method and the regression method. The regression method is used in the study to compute the factor scores. The positive zones indicate dominance of that factor (hydro geochemical regime). The spatial variation was estimated by using factor score values of each sampling points and they were plotted by Map Info Professional 10 (figure7)

The spatial distributions of dominant regions irrespective of all seasons are explained in five factors. Weathering is dominant in the NW part of the study area and the water table movement is from the NW to SE direction [36]. It indicates that weathering processes is dominant subsequently by anthropogenic activities. The higher concentration of U is distributed in the central and NE part of region. Higher concentration is mainly due to flushing of U by infiltrating rainfall recharge which results in increase of uranium in groundwater along with the rise in water level [37]. There are different studies of U in groundwater carried throughout the world in various types of granitic terrains and it is evident from the previous studies [38] that most of the regions with higher concentration of Uranium are found in the Granitic terrains of the world.

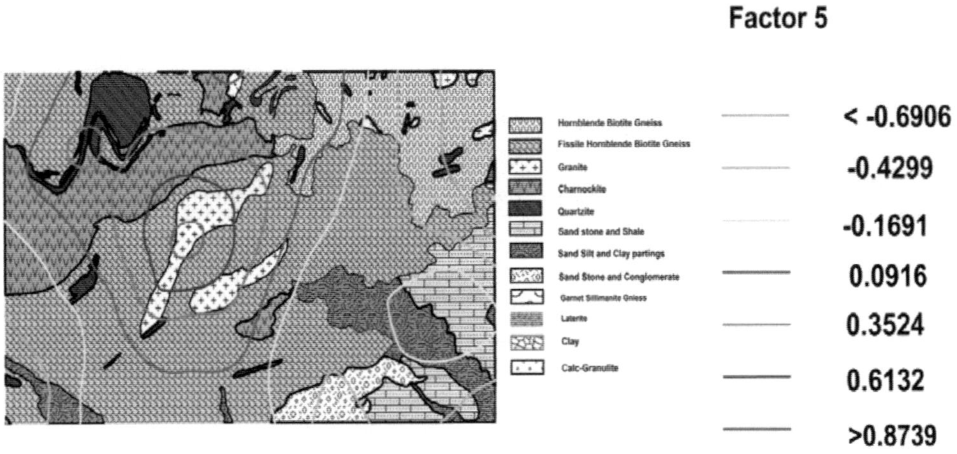

Figure 7. Overlay of Spatial distribution of uranium factor on lithology

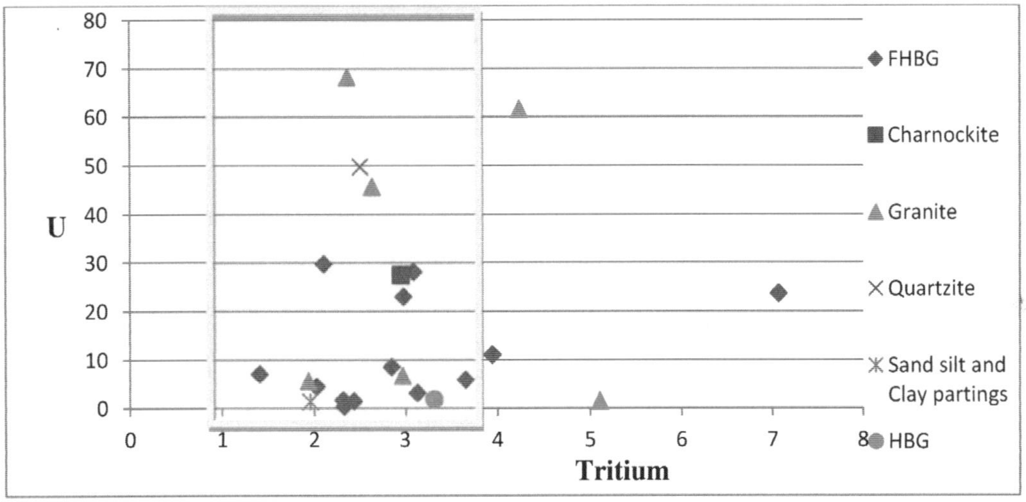

Figure 8. Scatterplot between total U and Tritium in groundwater samples

Tritium Conc.	Qualitative "Age" (afterClark and Fritz [39])
<0.8 TU	Sub modern—recharged prior to 1952
0.8 to ~4TU	Mixture between submodern and recent recharge
5 to 15 TU	Modern (<5 to 10 yrs)
15 to 30 TU	some "bomb" 3H present
>30 TU	Considerable component of recharge from 1960s or 1970s
>50 TU	Dominantly the 1960s recharge

In the study area based on Tritium values (figure 8)waters can be classified into 3 types such as sub modern, Mixture between sub modern and recent recharge and modern waters. Only selected samples were analyzed for Tritium. Most of the samples belonged to the mixture between sub modern and recent recharged type. In general highest uranium concentration is observed in this mixed water type belonging to rock types such as Granite, Fissile Hornblende Biotite Gneiss and Quartzite. The inference is found to match with that of Ionic strength which represents higher uranium values associated with hard rocks that have longer residence time for water and hence older waters with respect to tritium has higher uranium concentrations. It is found that modern waters contain lower uranium due to flushing.

5.Conclusion

The Uranium concentration in the present study had an average value of 11.78 ppb with a maximum of 68.28 ppb and 0.4ppb as the lowest reported value. The concentration in 95% of samples were less than 60 ppb - the AERB, Indian standard; whereas considering the WHO standards,88% of the samples were within the permissible limit of below 30 ppb of Uranium. Highest uranium concentration was noted in the central and NE side of the study area which comprises of hard rock terrain. Lithological dominance of Uranium concentration is in the following order; Granite reporting the highest U followed by Quartzite and Fissile Hornblende BiotiteGneiss. Few ground watersamples from Charnockite also showed high U concentration. From the Piper diagram it is concluded that major water type of the area belonged to Na-HCO_3 type responsible for higher Uranium. Few samples also belonged to Ca-HCO_3 type. Tritium results indicate that majority of water belonged to mixture between modern andsub modern water type and higher Uranium also corresponds to this water type. Thus from all these results it's clearly noted that sources responsible for higher uranium cannot be a single factor as seen in lithological influence; not only granites are responsible for higher concentrations. Similarly Ca- HCO_3 water types in the region also contain few higher Uranium values. Apart from all these sources responsible; redox potential is a major governing factor that contributes to higher Uranium in ground water.

References

[1] Siegel M D and BryanC R 2004 Environmental geochemistry of radioactive contamination Treatise on geochemistryp.262

[2] J. Michel 1991*Relationship of radium and radon with geological formations*Richard Cothern CR and Rebers PA (eds), Radon, Radium and Uranium in Drinking Water, 2nd ed, Lewis Publishers, Chelsea, MIp. 83–96.

[3] OrtegaX, Valles I and Serrano I 1996 Natural radioactivity in drinking water in Catalonia (Spain) *Environ Internat* vol.**22**, pp.347–54

[4] Ajay Kumar, Sabyasachi , Usha Narayanan , Manish K. Mishra , TripathiR M, Jaspal Singh, Sanjeev Kumar, and Kushwaha H S2011 Geochemical modelling of uranium speciation in the subsurface aquatic environment of Punjab State, India*Journal of Geology and Mining Research* Vol. **3(5)** p. 137-146

[5] Bucur C, Olteanu M, andPavelescuM2006Radionuclides diffusion in geological media. Romania," Journal of Physics, Vol. **51(3–4)** p. 469–478.

[6] Berzina I G, Yeliseyeva O P and Popenko D P 1975 Distribution relationships of uranium in intrusive rocks of northern Kazakhstan*Int. Geol. Rev*, Vol. **16(11)**p.1193-1204

[7] Muller M, Anke M and H 1997 Illing-GuntherAluminium in wild mushrooms and cultivated Agarciusbisporus.Z.Lebensm*Unters. Forsch.* Vol. **A 205**, p.242–247

[8] WHO (World Health Organization) Guidelines for drinking water quality recommendations, Geneva, 4th ed, Vol. 1, pp. 515, 2011.

[9] Central Ground Water Board, Annual report 2008-2009

[10] Hem J D 1985 Study and interpretation of the chemical characteristics of natural water *USGS water supply paper*, Vol. **2254**, pp.117–120

[11] Freeze A R and Cherry J A 1979*Groundwater*, Prentice – Hall, Inc Englewood cliffs, New Jersey, p. 604

[12] WHO (World Health Organization) Guidelines for drinking water quality recommendations, Geneva, 3rd ed, Vol. 1, pp. 515, 2006.

[13] AbdallaF Aand Scheytt T 2012 Hydrochemistry of surface water and groundwater from a fractured carbonate aquifer in the Helwan area. Egypt," Journal of Earth System Science, Vol. 121(1), p.109–124,

[14] Wayland K, LongD, HyndmanD, PijanowskiB, WoodhamsS and HaackK 2003 Identifying relationships between base flow geochemistry and land use with synoptic sampling and R-Mode factor analysis," Journal of Environmental Quality, Vol.**32** p.180–190

[15] Kortatsi B K Hydrochemistry of groundwater in the mining area of Tarkwa-Prestea, Ghana 2002Unpublished Ph.D. Thesis, University of Ghana, Legon-Accra p. 70–85.

[16] Anku S Y, Banoeng-Yakubo B and Yidana S M2008Water quality analysis of groundwater in crystalline basement rocks Northern Ghana Environmental Geology, Vol. **58(5)** 989–997. doi:10.1007/s00254-008-1578-4

[17] ChidambaramS, Karmegam U, PrasannaM V, Sasidhar P, and Vasanthavigar M 2011 A study on hydrochemical elucidation of coastal groundwater in and around Kalpakkam region, Southern India *Environmental Earth Sciences* Vol.**64**

[18] ChauN D andMichalec B 2009Natural radioactivity in bottled natural spring, mineral and therapeutic waters in Poland*JRadioanalNuclChem*p.121–129

[19] De CamargoI M C andMazzilli B1996 Determination of uranium and thorium isotopes in mineral spring waters*J RadioanNuclChem* Vol. **212(4)** p.251–258

[20] KilliariT andPashalidis I 2010 Simplified alpha-spectroscopic analysis of uranium in natural waters after its separation by cation- exchangeRadiatMeasurem Vol. **45(8)** p.966–968

[21] Stevens W, Bruenger F W,AthertonDR , Smith J Mand Taylor G N1980The distribution and retention of hexavalent 233 U in the beagle*Radiation Research* Vol. **83 (1)** 109–126

[22] Morrow P, Gelein R, BeiterH, Scott J, Picano J and YuileC1982 Inhalation and intravenous studies of UF6/UO2F2 in dogs*Health Phys* Vol. **43** 859-873

[23] Zamora M L, Tracy B L, and Zielinski J M 1998Chronic ingestion of uranium in drinking water: a study of kidney bioeffects in humans*ToxicolSci* Vol.**43** p.68–77

[24] Wrenn M E, N P Singh, and RuthH 1989 Gastrointestinal absorption of soluble uranium from drinking water by man*RadiatProtDosim* Vol. **26** p. 119–122

[25] Kurttio P, Auvinen A, .SalonenL, SahaH, Pekkanen J, Makelainen I, Vaisanen S B, Penttila I M, Komulainen H2002 Renal effects of uranium in drinking water," Environmental Health Perspectives **110** p.337-342

[26] SkeppstromK andOlofssonB 2007 Uranium and radon in groundwater. An overview of the problem*European Water***17/18** 51-62

[27] TempleT2002personal communication via email, June 14

[28] Ingo Braum 2006 Pan-African granitic magmatism in the Kerala Khondalite Belt, southern India*Journal of Asian Earth Sciences*Vol .**28(1)** p. 38-45

[29] Munazzam Ali, Aqeel Ahmed Shariff, Nasser Ali Qamar , andAmanullahLaghari 2011 An appraisal of uranium source potential of granites, associated felsic rocks, kaolin and calcretes of Nagar Parkar area, Tharparkar Pakistan *Journal of Himalayan earth sciences* Vol. **45(2)**

[30] LangmuirD1978Uranium solution-mineral equilibria at low temperatures with applications to sedimentary ore deposits *GeochimicaetCosmochimicaActa* Vol.**42(6)**547–569

[31] A. Ioannidou, I. Samaropoulos, Efstathiou M, and I .Pashalidis (2011) Uranium in ground water samples of Northern Greece.JRadioanalNuclChem, 289:pp.551–555

[32] Brasser T, Noseck U and SchönwieseD 2002*Uranium Migration in Argillaceous Sediments as Analogue for Transport Processes in the Far Field of Repositories* (Heselbach Site, Germany) in Merkel, B. J., Planer-Friedrich, B. &Wolkersdorfer, Ch.: Uranium in the Aquatic Environment p. 141-148

[33] ZapeczaO SandSzaboZ 1986 Natural radioactivity in ground water- a review. U.S. Geological Survey National Water Summary 1986, Ground- Water Quality: Hydrologic Conditions and Events *U. S. Geological Survey Water Supply Paper***2325**, p. 50–57

[34] Ivanovich M, Fröhlich K and Hendry M J1991Uranium-series radionuclides in fluids and solids, Milk River aquifer, Alberta, Canada *Appl. Geochem***6,** 405-418

[35] Chidambaram S, Bala Krishna Prasad M, Manivannan R, .KarmegamU,Singaraja C, Anandhan P, Prasanna M V andManikandan S 2012 Environmental hydrogeochemistry and genesis of fluoride in groundwaters of Dindigul district, Tamilnadu (India)*Environ Earth Sci.* doi:10.1007/s12665-012-1741-9,

[36] Thivya C, Chidambaram S, Thilagavathi R, Prasanna M V, Singaraja C, Nepolian M and Sundarrajan M 2013 Identification of the geochemical processes in groundwater by factor analysis in hard rock aquifers of Madurai district, South India. *Arab J Geosci*DOI 10.1007/s12517- 013-1065-4

[37] BrindhaK and Elango L2012Impact of tanning industries on groundwater quality near a metropolitan city in India," Water Resour. Manag, **26(6)**p.1747–1761

[38] Thivya C, Chidambaram S, TirumaleshKeesari, Prasanna M V, Thilagavathi R, Adithya V Sand Singaraja C 2015 Lithological and hydrochemical controls on distribution and speciation of uranium in groundwaters of hard-rock granitic aquifers of Madurai District, Tamil Nadu (India) *Environ Geochem Health*DOI 10.1007/s10653-015-9735-7

[39] Clark I and Fritz P 1997 *Environmental Isotopes in Hydrogeology*, Lewis Publishers, Boca Raton

Acknowledgements
The authors express their sincere thanks to Board of Research in Nuclear Sciences (BRNS), India, for providing the necessary financial support to this research project wide ref. No. 2012/35/12/BRNS/1918 dated 2012.The authors also thank to Curtin University, Sarawak Malaysia for the publication fee through Research Performance Index (RPI) fund of M.V Prasanna.

IOP Conf. Series: Materials Science and Engineering **121** (2016) 012010 doi:10.1088/1757-899X/121/1/012010

Cloud computing for energy management in smart grid – an application survey

P Naveen[1], Wong Kiing Ing[1], Michael Kobina Danquah[2], Amandeep S Sidhu[3, 4] and Ahmed Abu-Siada[5]

[1] Electrical and Computer Engineering Department, [2] Chemical Engineering Department, [3] Curtin Sarawak Research Institute, Curtin University Sarawak, Miri, Malaysia

[4] Faculty of Health Sciences, [5] Electrical and Computer Engineering Department, Curtin University, Perth, Australia

E-mail: naveen.p@postgrad.curtin.edu.my

Abstract. The smart grid is the emerging energy system wherein the application of information technology, tools and techniques that make the grid run more efficiently. It possesses demand response capacity to help balance electrical consumption with supply. The challenges and opportunities of emerging and future smart grids can be addressed by cloud computing. To focus on these requirements, we provide an in-depth survey on different cloud computing applications for energy management in the smart grid architecture. In this survey, we present an outline of the current state of research on smart grid development. We also propose a model of cloud based economic power dispatch for smart grid.

1. Introduction

Electrical power has become an indispensable part of modern day life. Hebra [1] and NIST [2] styled today's electric power system as a multifaceted system of power generation, transmission, and distribution. With the global economy more reliant on sustainable development of energy, a series of problems, such as energy shortage, electricity blackout and global warming are gaining attention. ABB [3] pointed out that there are tenacious economic as well as environmental urgings for the refurbishment of the conventional power systems, and its replacement with a Smart Electrical Power Grid or simply Smart Grid. A smart grid is an electricity network that uses digital and other advanced technologies to monitor and manage the transport of electricity from all generation sources to meet the varying electricity demands of end-users [4].

Energy demand from the users changes dynamically in different time-periods. The existing power grids need optimal balancing of electricity demand and supply between the customers and the utility providers. To address these requirements in smart grid, the energy management systems (EMS) such as building energy management (BEMS), demand side management (DSM) and home energy management (HEM) are integrated. A smart grid allows various renewable energy sources to have an efficient management of supply and demand. The special characteristic of a smart grid is its heterogeneous architecture, which includes Demand Response (DR), distributed generation, resource scheduling, and real-time pricing model.

In a smart grid environment, multiple devices are implemented such as smart meters, substations, micro-grids, home appliances, sensor nodes, and communication network devices. Smart meters, deployed in the distribution sites, generate massive data for real-time communication with the utilities. To handle such massive data efficiently and effectively, smart grid relies on the utilization and integration of advanced information technologies [5]. One of the important trends in today's information management is outsourcing tasks to cloud computing, which has been regarded as the next-generation computing and storage paradigm [6].

1.1. Contribution
In this paper, we provide a methodical summary of assimilating cloud computing applications for energy management in smart grids. Communication network plays an important role for reliable energy management as smart grid is the combination of electrical and communication network. Cloud computing techniques proposed in the existing literature for energy management in smart grid are discussed briefly. A model of power dispatching in smart grid with cloud is described. In short, our objective in this paper is to offer the following:

- A comprehensive overview of smart grid and cloud computing in terms of energy management.
- A clear concept of cloud computing applications in smart grid.
- A highlight on cost effective cloud based power dispatching in smart grid.

The rest of the paper is organized as follows. In Section 2 and Section 3, we give a comprehensive overview of smart grid and cloud computing. We briefly describe the cloud computing applications in the context of energy management in smart grid in Section 4. Cost effective cloud based power dispatching in smart grid is presented in Section 5. Finally, a conclusion is provided in Section 6.

2. Smart Grid
Smart Grid monitors power use, adapts consumption to match power costs and system load, and integrates new kinds of renewable energy sources with conventional power generating systems. It associates every distributed electricity producer (independent power producer) in the energy market from conventional thermal, hydroelectric and atomic power plants to new kinds of renewable energy systems with each electricity consumer, from industries to residences, and to every load plugged into the electric network. The digitally controlled, self-monitoring and self-healing Smart Grid provides two-way communication for energy production, transmission and distribution, control and monitoring, supply and demand balancing, etc. with more customer choices. According to NIST [2] "a modernized grid that enables bidirectional flows of energy and uses two-way communication and control capabilities that will lead to an array of new functionalities and applications". The NIST Smart Grid Framework is shown in figure 1, which shows the involved domains, the actors in a particular domain and interaction between the actors from other domains. Each domain encompasses Smart Grid actors and applications. Actors make decisions and information exchange that are required for performing applications. Actors include devices (e.g. smart meters, solar energy generators, and control systems), programs, systems, and stakeholders. Tasks performed by one or more actors within a domain are called applications (home automation, and energy management).An important characteristic of a smart grid is controlling electricity consumption at the customers' ends [7]. To achieve this metering and monitoring, a few components are incorporated in the smart grid architecture. These key components of smart grid are discussed as under:

2.1. Advanced Metering Infrastructure
Advanced metering infrastructure (AMI) is an integrated system of smart meters, communications networks, and data management systems that enables two-way communication between utilities and customers [8]. Smart meters at the customer site are the electrical meters that record real-time readings i.e. consumption of electrical energy, and voltage quality that are anticipated to be read

periodically in shorter intervals (range from minutes to milliseconds). In general, AMI support data communication architecture by including software and electronic/digital hardware.

2.2. Advanced Metering Infrastructure

Health and performance monitoring of transformers, feeders, capacitors, circuit breakers, condition monitoring of lines (e.g. PLCs), fault detection and replacement of assets can be performed effectively and efficiently by the analysis of data sensed at regular intervals by the intelligent monitoring system. With the development of sensors that utilize low-cost, ultra-low-power processors and communication links to transmit data, a new regime of sensors known as the smart sensors have recently come into existence [9]. Extensive research in the wireless technology reducing the signal noise, power requirement and enhancing the range has made the wireless technology a potential for integration with smart sensors to carry communication with utilities. Development of effective network topologies such as multi-hop link networks and efficient wireless signal routing algorithms have further improved the range and reduced power consumption; hence made the advent of protocols like ZigBee Pro [10]. Some examples of sensors present in the commercial domain are Tollgrade's Lighthouse MV sensors, USi's Power Donut, ABBs Grid Sync, Grid Sense's Line IQ, Sentient monitor, and Grid Sentry's Line sentry. The size, cost and weight restrictions of the sensors mentioned above have posed the hindrance to their wide acceptance.

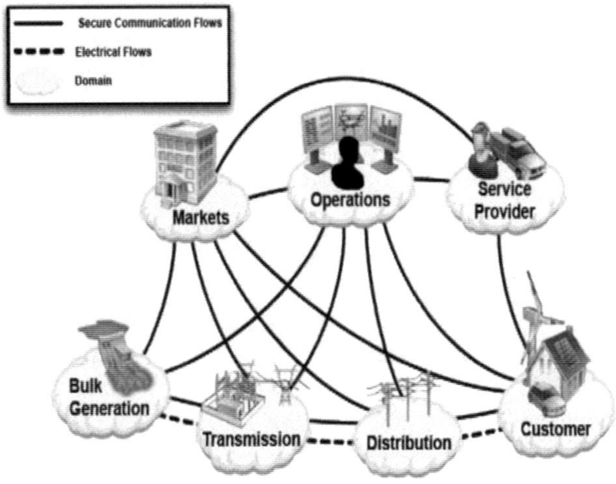

Figure 1. NIST Smart Grid Framework.

3. Cloud Computing

Cloud computing - a model that enables convenient, ubiquitous, on-demand access to a pool of computing resources (e.g. servers, networks, applications, storage, and services) that are configurable. With minimal management effort, resources can be provisioned and released seamlessly [11]. NIST's visual model of cloud computing definition is shown in figure 2. It delivers infrastructure, platform, and software to customers as subscription-based services in a pay-as-you-go model. The advantages, essential characteristics, of using a cloud computing model are as follows [12-14]

- On-demand self-service: A consumer can individually provision computing capabilities as needed automatically without requiring human interaction with each service provider.
- Broad network access: Capabilities are available over the network. It can be accessed through standard mechanisms, to be used by heterogeneous thin or thick client platforms.

- Resource pooling: A multi-tenant model is used to serve multiple consumers from a pool of computing resources. The customer has no control over the exact location of the provided resources.
- Rapid elasticity: Cloud computing supports elastic nature of storage and memory devices. It can expand and reduce itself according to the demand from the users, as needed.
- Measured service: Cloud computing offers metering infrastructure to customers. Cost optimization mechanisms are offered to users, enabling them to provision and pay for their consumed resources only.

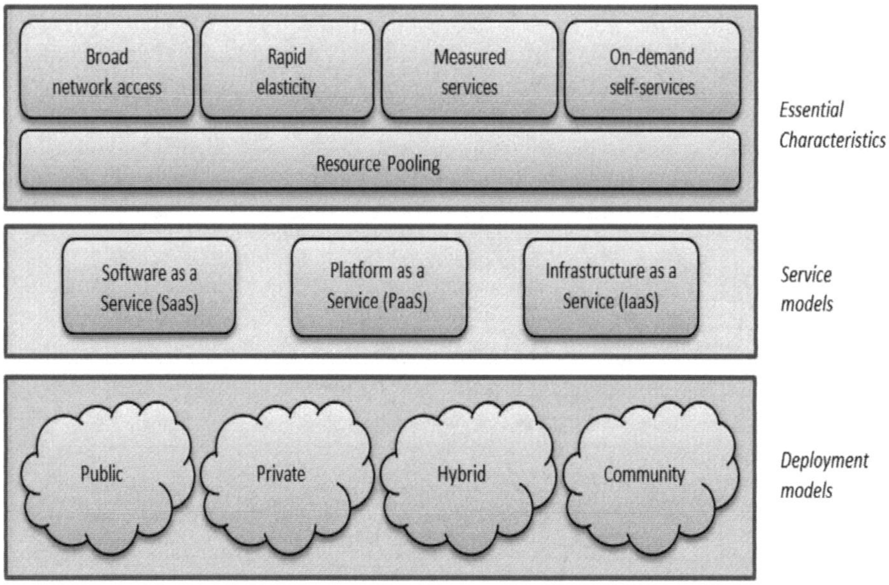

Figure 2. NIST Visual model of cloud computing definition.

Virtualization technology can be used in cloud computing that can take a variety of different types of computing resources as abstracted services to users. These cloud services are divided into Infrastructure-as-a-Service (IaaS), Platform-as-a-Service (PaaS) and Software-as-a-Service (SaaS).

- Infrastructure-as-a-Service: IaaS provides scalable infrastructure e.g. servers, network devices, and storage disks to consumers as services on demand. The access to the cloud is provided through various user interfaces, such as web service application programming interface (API), command-line interfaces (CLI) and graphical user interfaces (GUI) which provide different level of abstraction. The consumer has control over operating systems, storage, and deployed applications but they are not required to manage or control the cloud infrastructure.
- Platform-as-a-Service: PaaS provides a platform where users or customers can create and run their applications or programs. The users can build and deliver Web-applications without downloading and installing required software, as PaaS service completes the requirements. It is responsible for the runtime execution of users' given task. The most important customers for this layer are the developers.
- Software-as-a-Service: SaaS is responsible for delivering various kinds of applications plus the interfaces for the end users. This feature of cloud computing is accessible through Web-browsers. The SaaS provides the modeling of software deployment where users can run their

applications without installing software on his/her own computer. However, this service is limited to the users, i.e., only existing set of services is available to the customers.

According to the deployment model, a cloud can be classified as public cloud, private cloud, community cloud, and hybrid cloud[11].

- Private cloud: The Private cloud is a virtual environment deployed within an organization that is restricted to users within the company and usually resides behind the corporate firewall. It is suited for secured confidential information and core systems. It may exist on or off premises.
- Community cloud: The Community cloud is similar to a Private cloud. It is provisioned to a group of organizations who have similar type of requirements with additional features. It may exist on or off premises.
- Public cloud: The Public cloud is a virtual environment that is publically available for any consumer to purchase computing resources, usually on a pay per use basis, via an easy to use web portal. It exists on the premises of the cloud provider.
- Hybrid cloud: The combination of the Public and Private cloud whereby specific resources are used in Public Cloud while others are used in Private Cloud.

4. Cloud Applications for Energy Management

Energy management is the process of monitoring, controlling, and conserving energy [15]. In smart grid, energy management is a major concern [16]. It is needed for resource conservation, climate protection and cost saving without compromising work processes by optimally coordinating several energy sources. BEMS (Building Energy Management System) and HEMS (Home Energy Management System), dynamic pricing, and load shifting are different applications that are implemented by researchers in the past to address energy management.

4.1. Problems with Existing Approaches without cloud

Demand Response (DR) refers to "changes in electric usage by end-use customers from their normal consumption patterns in response to changes in the price of electricity overtime, or to incentive payments designed to induce lower electricity use at times of high wholesale market prices or when system reliability is jeopardized" [17, 18]. DR is achieved through the application of a variety of DR resource types, including distributed generation, dispatchable load, storage and other resources that may contribute to modify the power supplied by the main grid [19]. In the conventional smart grid architecture (without cloud), several problems, as detailed below, are addressed by researchers [12, 13, 20, 21].

- Master-Slave architected (without clouds) could cause Cyber-attacks (Distributed Denial of-Service (DDoS)).
- Any failure in Master-Slave architecture could lead to a system failure, which does not exist in cloud computing.
- Can serve for limited number of users (customers) due to limited server capacity.
- Serving of such large number of the customer will be challenging because of limited memory and storage.
- Management, as well as stability issues, will be required.

4.2. Solution Concept with Cloud Applications

For many years, researchers proposed several solution concepts for demand response and micro-grid management [12-14, 20, 22-39].

Kim et al. [12] proposed the concept of Cloud-Based Demand Response (CDR) for fast response times in large scale deployment. In the concept of CDR, the Energy Management System (EMS) and smart meters will be the slaves while master will be the utility. The data-centric communication, publisher/subscriber will be used by the CDR, whereby two cloud-based demand models are suggested (a) data-centric communication and (b) topic-based group communication rather than typical IP-centric communication. Overhead problems, such as implementation cost, and the selection of appropriate strategy, exists in the demand response model [12] that occurs in the private cloud when the size of the network is small [7].

Ming Chen et al. [22] analyzed the necessity and feasibility of cloud computing technology in power dispatching and presented the Deployment Method of power dispatching automation system based on cloud computing. Easy standardization of power dispatching technique; rapid delivery of advanced functions; and significant improvement reliability of IT infrastructures can be achieved by means of cloud computing technology. It reduces administrative costs, and it solves the contradiction between hierarchical management and "integrated construction" in energy sector [22].

ZHANG Liang [23] proposed the concept of cloud dispatching, a kind of cloud computing-based overall framework of intelligent dispatching center. The cloud computing-based layered architecture includes Physical Resource Layer- physical hardware, platform resources and application systems such as SCADA, EMS, TMR, WARMS, Virtual Resource Layer - map physical resources in various types into virtual resources, Cloud Service layer - packages virtual resources into services, that are posted to the clouds, Cloud Management Layer - provides integrated management of cloud services for users, and Cloud Access Layer - the way user access cloud dispatching. It integrates the existing resources demands among various dispatching centers, reduces the system construction and expansion cost and improves overall dispatching business ability.

The smart grid infrastructure needs to be deployed globally. In order to balance the real-time demand and supply curves, rapid integration and analyzation of information that streams from multiple smart meters simultaneously is required that necessitates the scalable software platform. Yang et al. [13] proposed that cloud platforms are well suited to support huge data and computationally-intensive, always-on applications. To build a software infrastructure to support such dynamic and always-on applications, scalable requirements of resources are offered by the cloud applications. In these environments, cloud platforms serve as essential components due to the various benefits they offer, as mentioned below:

- Cloud acts elastically to avoid costly capital investment by the utility during the peak hours.
- Customers can be benefited from the real-time information by sharing the real-time energy usage and pricing information.
- Some data can be shared with a third party by using cloud services, after meeting the data privacy policies for developing intelligent applications to customize consumer needs.

In order to take decisions at different instances, implementation of specialized data abstraction for data streams generated from the different components is required for real-time monitoring. On the other hand, third-party vendors are allowed to participate in such real-time monitoring system that necessitates defining an effective privacy policy as a security mechanism.

Virtualization is one of the most efficient techniques for cost reduction, resource optimization, and server management [20]. Cloud computing can be implemented in the form of different strategies of the micro-grids. Rajeev and Ashok [20] proposed a framework for integrating cloud computing applications for micro-grid management in the form of different modules such as infrastructure, power management, and service. The infrastructure and power management modules are used for task scheduling and micro-grid power management respectively. The different operators publish their service description using the service module. With the implementation of cloud computing, the external computing devices can be integrated with the internal ones. Thus, the number of supported customers increases as suggested by Rajeev and Ashok [20]. In such a manner, integrating virtual energy sources with the existing energy storage devices and the energy exchange mechanism can be achieved among the micro-grids to meet the energy requirements from consumers.

Dynamic pricing implementation can be used to address energy management. Xuan Li and Lo [14] proposed two smart grid related issues: (a) peak demand and (b) dynamic pricing. Requests from customers, which are to be executed, based on the priority, available resources, and other applicable constraints, are scheduled with the integration of cloud. During peak hours, the messages from smart meters are more than those in the nonpeak hours [14]. In such a scenario, incoming jobs from users are scheduled according to their priority, available resources, and applicable constraints. With the integration of dynamic bandwidth allotment mechanism using cloud application, these issues can be addressed conveniently. During the peak-hour, the allotted bandwidth is higher than that in the non-peak hour, in order to serve all the incoming jobs simultaneously.

Cloud-based services are used for communication and management schemes in the smart grid by Ji et al. [24], while providing the facility of power monitoring and early-warning system as well. In such a scenario, the real-time support is provided by using enterprise service bus (ESB) and service-oriented architecture (SOA). On-demand, efficient, flexible, and scalable smart grid power monitoring system can be built by this approach. Standard Web services, service finding, service registry, interfaces, and service access are implemented into a single cloud application using SOA, as the SOA relies on publishing applications as a service. Smart grid energy management can be performed using the ESB architecture, which includes activities such as security management, task management, and resource management, with the implementation of cloud applications [24].

The Energy Service Interface (ESI) interconnects internal customer energy resources and external systems. Eun-Kyu Lee et al. [25] built an ESI test-bed which includes ESI, a demand response service server, and customer energy resources. A demand response client module is implemented by ESI. The customer energy data are represented by XML format and web service interfaces are implemented for inter-domain communications. An additional test-bed, in which the ESI is deployed on a public cloud. The two test-beds are deployed to verify that the ESI plays the service "prosumer" in a practical manner for a couple of energy service scenarios. The Energy Service Interface (ESI) interconnects internal customer energy resources and external systems.

The trend of the power grid to shift to the Smart Grid leads to the enormous pool of computing and massive data storage requirements. To overcome these demands of cloud computing, highly distributed and scalable computing resources, to host the smart grid applications was proposed. Bitzer and Gebretsadik [26] discussed the feasibility of handling the monitoring of renewable energy in smart grid on cloud computing framework by a Lab-demonstrator. The Lab-demonstration set up considers the power system and the cloud computing domains. The distributed energy resources and the local SCADA control are considered as the power system domain whereas the cloud computing domain contains a specific cloud computing provider [27]. Generation plants can be monitored and controlled by the local SCADA software running on the local computer. Through web-based SCADA monitoring and controlling of the plants can be done from anywhere.

Jinsung Byun et al. [28] proposed intelligent cloud home energy management system (iCHEMS) to address inefficient home energy management systems due to intermittent energy generation by renewable energy. In iCHEMS, a household appliance is assigned with dynamic priority according to its type and current status. The use of household appliances is scheduled based on the assigned priority and renewable energy capability. To enhance utilization of computing and renewable energy system resources, iCHEMS exploits cloud computing. Energy distribution, situation-based energy management, and user-centric energy management are services provided by cloud resources. With the help of cloud computing, the proposed system reduces the high cost, required to implement smart green home and average total power consumption.

Aras Sheikhi et al. [29] proposed a model for utilizing the cloud computing technology in the smart grid domain and explores how cloud computing plays an effective role in DSM (Demand Side Management) game among a group of Smart Energy Hubs (S.E.). Interaction between the utility company and demand response realizing functions, the amount of loads per customer are to be reduced and at which incentive price, are performed on the cloud. In this model, to reach an optimal DSM, based on the game theoretic approach, load profiles of S.E. Hubs are communicated to the CC (Cloud

Computing) where each S.E. Hub attempts to minimize its own energy cost in response to the aggregated information on the actions of the other users [30]. The result of the game, Nash Equilibrium (NE), leads to a proper strategy for each S.E. Hub to minimize their energy bill [29]. The DSM game reduces the PAR (Peak to average ratio) in the electricity grid and the daily energy charges of each S.E. Hub can be significantly reduced.

Rajeev and Ashok [31] proposed a dynamic load-shifting program, makes use of real-time data, in a cloud computing framework to address the forecasting and operational challenges issues, which are to be met when implementing effective renewable energy program. A new dynamic renewable factor, a reference parameter (captures and represents the dynamics in the pricing and shifting strategy) in the algorithm, is proposed to facilitate on-time incentive based load shifting program. With the help of cloud-based infrastructure, the widely distributed renewable energy sources operations are coordinated by the utility at a minimal cost. In addition improved utilization renewable sources, reduction in the peak demand at domestic level, additional household annual bill savings are the benefits of dynamic shifting program.

Smart grids with the usage of information technologies enable efficient power grid. To cope with the huge amount of data and daily fluctuations by the smart meters the underlying infrastructure must be (i) massively scalable and (ii) elastic. Cloud computing is a cost-efficient alternative to dedicated data centers. Martin et al. [32] explored the combination of an elastic Event Stream Processing (ESP) system named Stream Mine3G and cloud technologies in the context of energy forecasting. ESP aim at processing high volume streams of data by processing data on-the-fly instead of storing it first. StreamMine3G, scalable and elastic ESP, is equipped with a resource manager, acquires and releases virtual machines, enables load shifting from overloaded nodes to less loaded ones through migration of stateful and stateless operators. The elasticity properties of ESP system was showcased by performing several experiments on deployed StreamMine3G at Amazon EC2.

Neeraj Kumar et al. [33] proposed a context-aware layered architecture for demand side management using vehicular cyber–physical system (VCPS) with cloud support. With the integration of vehicles with cloud computing, storage, sensing, software, platform, and Network-as-a-Service (NaaS) are offered to the clients. Bayesian coalition game concept and learning automata, an intelligent context-aware data collection and processing, are used. In this scheme, players are the mobile vehicles, which can sense the SG environment and collect information from it. Alert generation and information dissemination are the actions performed by the players in the game. The player's action probability vector is updated based on the feedback from the environment to the players. Reduction in the energy shortage and information processing delay, the increase in energy sold back to the grid are achieved in the proposed scheme.

The importance of BEMS (Building Information Management System) is increasing as Smart Grid spreads. BEMS of each entity is improved to provide high-quality services. The number of the entity increases according to the coverage area of a building. Insung Hong et al. [34] proposed the Cloud computing-based BEMS to lessen the burden of each entity by a System Manager, a centralized server. To monitor power consumption and environmental information, sensor entities with 8-bit microcontroller and ZigBee and Low-cost power are used. The BEMS consists of three the Power Monitoring Entity (PME) - monitor and control device's power consumption, Environmental Information Entity (EIE) - collect environmental information, and System Manager - collect and manage data and provide services to users. This system reduces each entity's cost and hardware specification.

Effective DEM depends on load and renewable production forecasting that leads to large volumes of data generated by a vast number of smart meters. In order to optimize the smart grid operations, DEM requires high performance computing, efficient data network management, robust data analytics, and cloud computing techniques [35]; the cloud computing model meets these requirements [36]. In Literature, approaches have been developed to increase the energy efficiency of High Performance Computing (HPC) data centers, such as the cooperation with the SG in [37], energy conscious scheduling in [38]. Most of smart grid applications, such as advanced metering infrastructure, SCADA, and energy management can be facilitated by the available cloud service models, namely software as a service, platform as a service, and infrastructure as a service [39].

5. Cloud Based Economic Power Dispatching

Smart grids needs to be equipped with an integrated solution to the problem of modern energy delivery network that enables two-way energy and communication with the customers. To manage large amounts of data, cloud computing is the best way for smart grids due to its scalable, economical, and flexible characteristics. We propose a cloud based dispatching model in the smart grid domain. The prime responsibility of the Electric Utilities is to meet the customers' requirements at all times with quality and quantity as agreed. It matches the power generation by the utilities against the customers' power demand at all times. As the consumers' demand changes at every instant, the power generation by the Utilities should match with the consumers' demand. In reality, the power generation can't be adjusted at every instance; hence the generation is adjusted normally at 20 minutes interval. The matching of power generation against the consumers' demand is known as Power Dispatching (Hongseok Kim et al. [12], and Palanichamy et al. [40]). In cloud based economic power dispatching model, the utility and customers interact through the cloud, and the functions for cost optimization are performed in the cloud. From utility's perspective, cloud appears to be an information system, which takes an input from utility (e.g., Power demand, weather data, fuel cost etc.), processes the information, and gives an output to utility and customers (e.g., generations of the individual plants, total production cost etc.). It is worth pointing out that the Cloud network performs the power dispatching job as per the instruction of the Electric Utilities since the decision authority is the Utility. For its services, the Cloud gets it service charges from the Utilities. This arrangement is economical for the Utilities since they need not invest on communication and computing facilities.

6. Conclusion

In this survey, we provided an overview of existing works integrating cloud computing in the existing smart grid architecture, in order to have a reliable and efficient energy distribution. Different aspects of energy management in the smart grid were discussed. We identified some important technical issues and proposed several future research directions on cloud-based smart grid. Using cloud computing applications, energy management techniques in smart grid can be evaluated within the cloud, instead of between the end user's devices. This architecture gives more memory and storage to evaluate computing mechanism for energy management, and cost-optimization. We proposed a new highlight on cost-effective cloud based power dispatching for smart grid applications. From this surveyed work, we can see the integration of cloud computing in smart grid is envisioned to be useful for evolving the smart grid architecture further in terms of considerations such as monitoring cost, computing, and power management.

References

[1] Hebra 2012 The reality of generating and transmitting ultrahigh voltage power Part 1 *IEEE Instrumentation & Measurement Magazine* **15(2)** 26–31
[2] Locke G and Gallagher P D 2010 *NIST framework and roadmap for smart grid interoperability standards release 1.0* (USA:National Institute of Standards and Technology) p 33
[3] ABB 2009 *Toward a smarter grid ABB's vision for the power system of the Future* (USA: ABB Inc.)
[4] IEA 2011 *Technology roadmap smart grids* (France: International Energy Agency)
[5] Fang X, Misra S, Xue G and Yang D 2012 Smart grid - the new and improved power grid: a survey *IEEE Communications Surveys & Tutorials* **14(4)** 944–80
[6] Xi F, Dejun Y and Guoliang X 2013 Evolving smart grid information management cloudward: a cloud optimization perspective *IEEE Transactions on Smart Grid* **4(1)** 111–9
[7] Bera S, Misra S and Rodrigues J J 2015 Cloud computing applications for smart grid: a survey *IEEE Parallel and Distributed Systems* **26(5)** 1477–94

[8] SMARTGRID.GOV 2015 *Advanced metering infrastructure and customer system* (USA: U.S. Department of Energy)

[9] Ukil 2008 Towards networked smart digital sensors: a review *Proc. 34th Annual Conf. of IEEE on Industrial Electronics (IECON) (Orlando)* (USA: IEEE)1798–802

[10] Barrenetxea G, Ingelrest F, Schaefer G and Vetterli M 2008 Wireless sensor networks for environmental monitoring: the sensorscope experience *Proc. Int. Seminar on Communications (Zurich)* (USA: IEEE) 98–101

[11] Mell P and Grance T 2011 *The NIST definition of cloud computing* (USA:National Institute of Standards and Technology)

[12] Kim H, Kim Y J, Yang K and Thottan M 2011 Cloud-based demand response for smart grid: architecture and distributed algorithms *Proc. Int. Conf. on Smart Grid Communications (SmartGridComm) (Brussels)* (USA: IEEE) 398–403

[13] Yang C T, Chen W S, Huang K L, Liu J C, Hsu W H and Hsu C H 2012 Implementation of smart power management and service system on cloud computing *Proc. 9th Int. Conf. on Ubiquitous Intelligence & Computing and Autonomic & Trusted Computing (UIC/ATC) (Fukuoka)* (USA: IEEE) 924–9

[14] Li X and Lo J C 2012 Pricing and peak aware scheduling algorithm for cloud computing *Proc. Innovative Smart Grid Technologies (ISGT) (Washington, DC)* (USA: IEEE) 1–7

[15] Energy lens *Energy Management* (UK: BizEE Energy lens)

[16] Marta Marmiroli 2014 *Developing and testing a next generation energy management system* (USA: IEEE Smart Grid)

[17] QDR Q 2006 *Benefits of demand response in electricity markets and recommendations for achieving them* (USA: U.S. Department of Energy)

[18] Chiu, Ipakchi A, Chuang A, Qiu B, Brooks D, Koch E, Zhou J, Zientara M, Precht P and Burke R 2009 *Framework for integrated demand response (DR) and distributed energy resources (DER) models* NAESB & UCAIug

[19] Siano P 2014 Demand response and smart grids - a survey *Renewable and Sustainable Energy Reviews* **30** 461–78

[20] Rajeev T and Ashok S 2011 A cloud computing approach for power management of microgrids *Proc. Innovative Smart Grid Technologies-India (ISGT India) (Kollam, Kerala)* (USA: IEEE) 49–52

[21] Soultanis N L, Papathanasiou S A and Hatziargyriou N D 2007 A stability algorithm for the dynamic analysis of inverter dominated unbalanced LV microgrids *IEEE Transactions on Power Systems* **22** 294–304

[22] Chen M, Bai X Zhu Y and Wei H 2012 Research on power dispatching automation system based on cloud computing *Proc. Innovative Smart Grid Technologies-Asia (ISGT Asia) (Tianjin)* (USA: IEEE) 1–6

[23] Liang Z, Lei Z, Min-hui G and Xiao-liang B 2011 Research and application on the cloud-computing-based power dispatching IT architecture *Proc. Power and Energy Engineering Conference (APPEEC) Asia-Pacific (Wuhan)* (USA: IEEE) 1–4

[24] Ji L, Lifang W and Li Y 2012 Cloud Service based intelligent power monitoring and early-warning system *Proc. Innovative Smart Grid Technologies-Asia (ISGT Asia) (Tianjin)* (USA: IEEE) 1–4

[25] Lee E K, Gadh R and Gerla M 2013 Energy service interface: accessing to customer energy resources for smart grid interoperation *IEEE Journal on Selected Areas in Communications* **31(7)** 1195–1204

[26] Bitzer and Gebretsadik E S 2013 Cloud computing framework for smart grid applications *Proc. 48th Int, Universities' Conf. on Power Engineering (UPEC) (Dublin)* (USA: IEEE) 1–5

[27] Bitzer and Gebretsadik E S 2014 Cloud computing for monitoring and controlling of distributed energy generations *Proc. 49th Int. Universities' Conf. on Power Engineering (UPEC) (Cluj-Napoca)* (USA: IEEE) 1-5

[28] Byun J, Hong I and Park S 2012 Intelligent cloud home energy management system using household appliance priority based scheduling based on prediction of renewable energy capability *IEEE Transactions on Consumer Electronics* **58(4)** 1194–201

[29] Sheikhi, Rayati M, Bahrami S, Ranjbar A M and Sattari S 2015 A cloud computing framework on demand side management game in smart energy hubs *Int. Journal of Electrical Power & Energy Systems* **64** 1007–16

[30] Sheikhi, Rayati M, Bahrami S and Mohammad Ranjbar A 2015 Integrated demand side management game in smart energy hubs *IEEE Transactions on Smart Grid* **6(2)** 675–83

[31] Rajeev T and Ashok S 2015 Dynamic load-shifting program based on a cloud computing framework to support the integration of renewable energy sources *Applied Energy* **146** 141–9

[32] Martin, Silva R, Brito A and Fetzer C 2014 Low cost energy forecasting for smart grids using Stream Mine 3G and Amazon EC2 *Proc. IEEE/ACM 7th Int. Conf. on Utility and Cloud Computing (UCC) (London)* (USA: IEEE) 523–28

[33] Kumar N, Singh M, Zeadally S, Rodrigues J J and Rho S 2015 Cloud-assisted context-aware vehicular cyber-physical system for PHEVs in smart grid *IEEE Systems Journal* **(99)** 1–12

[34] Hong, Byun J and Park S 2012 Cloud computing-based building energy management system with zigbee sensor network *Proc. 6th Int. Conf. on Innovative Mobile and Internet Services in Ubiquitous Computing (IMIS) (Palermo)* (USA: IEEE) 547–51

[35] Diamantoulakis P D, Kapinas V M and Karagiannidis G K 2015 Big data analytics for dynamic energy management in smart grids *Big Data Research* **2** 94–101

[36] Rusitschka S, Eger K and Gerdes C 2010 Smart grid data cloud: a model for utilizing cloud computing in the smart grid domain *Proc. First IEEE Int. Conf. on Smart Grid Communications (SmartGridComm)* (*Gaithersburg, MD*) (USA: IEEE) 483–8

[37] Ghamkhari M and Mohsenian-Rad H 2013 Energy and performance management of green data centers: a profit maximization approach *IEEE Transactions on Smart Grid* **4(2)** 1017–25

[38] Lee Y C and Zomaya A Y 2011 Energy conscious scheduling for distributed computing systems under different operating conditions *IEEE Transactions on Parallel and Distributed Systems* **22(8)** 1374–81

[39] Markovic D S, Zivkovic D, Branovic I, Popovic R and Cvetkovic D 2013 Smart power grid and cloud computing *Renewable and Sustainable Energy Reviews* **24** 566–77

[40] Palanichamy, Naveen P and Muthusundar S 2014 A sustainable renewable energy mix option for the secluded society *Journal of Renewable Sustainable Energy* **6(2)** 023124

IOP Conf. Series: Materials Science and Engineering **121** (2016) 012011 doi:10.1088/1757-899X/121/1/012011

Solar power for energy sustainability and environmental friendliness of Curtin University Sarawak

A/Prof Dr C Palanichamy and Alvin Goh

Electrical and Computer Engineering Department, Curtin University Sarawak, Miri, Malaysia

E-mail: drcpc@curtin.edu.my

Abstract. The demand on electrical energy is rapidly increasing. Everything around us requires electrical energy either during its production or usage stage. Sustainability has become the main concern nowadays as the availability of fossil fuels is limited. As renewable energy is the path-way to energy sustainability and environmental friendly environment, this paper proposes a solar power system for Curtin University Sarawak to reduce its electricity consumption and greenhouse gas emissions. The proposed 208 kW solar system saves an energy consumption of more than 380,000 kWh per year, and a CO2 offset by 285 Tons per year

1. Introduction

Virtually every single product around us has gone through these stages as illustrated in figure 1: extraction, production, distribution, consumption and disposal [1, 2]. Each of these stages necessitates the use of electrical energy. With the ever-increasing rate of material productions, electrical energy demand is rising swiftly.

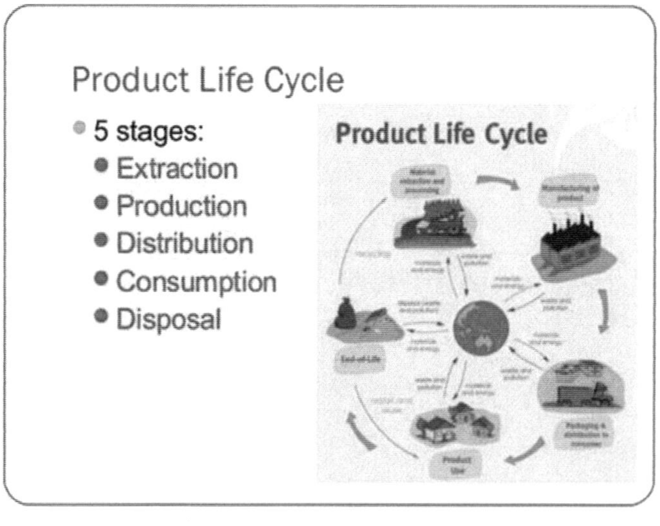

Figure 1. Product Life Cycle.

Extraction is a process when natural resources are harvested to be used as raw material. With the increased in extraction process, natural resources will soon be used up. During the past three decades, about 33 per cent of the planet resources have been consumed [3, 4]. In production stage, electrical energy is primarily used to convert extracted raw materials into products. Some production involves chemicals, which is harmful to health and the environment. Once products have been produced, they are distributed to stores for consumption. The rate of consumption is rising and the products consumed are not being used efficiently. Studies [1-5] have shown that most products are only being used for six months before being disposed. The end of the products life cycle is the disposal stage. The per capita garbage generation is around 0.45 to 0.60 kg and this figure will reach a level of 1.20 to 1.50 kg in the next decade [1, 6, 7]. Garbage is being disposed in a landfill or being burned before being disposed in the landfill. This process pollutes the environment as many products are produced through chemical process. Burning these products emits harmful toxics such as dioxin into the environment. Hence the disposal is not environmental friendly and involves energy consumption also. If the product life cycle continues like this without adequate social and economic concerns, then human safety, energy resources availability and environmental friendliness will become horrendous. The proven way for energy sustainability and environmental friendly atmosphere is introducing renewable energy sources at all sectors and reducing the dependency on depleting fossil fuels [8-13]. Renewable energy such as solar power is an alternative way to conventional approach of generating electricity [14-17]. Using renewable energy to generate electricity is a step towards a cleaner future.

Malaysia plans to achieve 985 MW or 5.5% share of renewable energy in the energy mix by 2015. By 2020, the target is for renewable energy to comprise 11% or 2,080 MW of overall electricity generation in the country. However, as to-date, renewable energy has yet to be considered a major source of electricity generation as it contributes less than 1% to the energy mix in Malaysia [18-20]. In order to support the Government's vision, Curtin University, Sarawak Malaysia (Curtin Sarawak) [21], the first and largest offshore campus of Curtin University in Perth, Western Australia and the first foreign university to be established in East Malaysia in partnership with the Sarawak State Government has exposed intense interest in presenting renewable energy systems. As sustenance to Curtin's concern, this paper intends a 208 kW solar power plant.

2. Solar Insolation

Curtin Sarawak is situated in Miri. Miri is a coastal city in northeastern Sarawak, Malaysia, located near the border of Brunei, on the island of Borneo. The geographic coordinates of Miri, Malaysia are [22]:

Latitude: 4°24′53″ N

Longitude: 114°00′32″ E

Elevation above sea level: 8 m

For the Curtin campus, the average peak sun hours for a twelve month period is calculated and presented in table 1. The peak sun hours are the number of hours in a day where the solar irradiation is 1000 W/m2. For Curtin, it varies from a minimum of 4.16 to a maximum of 5.55 hours during a year. Also the monthly insolation on unit area at an angle of 21° is calculated as shown in the table 1.

Table 1. Monthly Insolation.

Month	Peak Sun Hours Per Day (Hours)	Days/ Month	Monthly Insolation kWh/m2
January	4.98	31	262.82
February	4.56	28	232.70
March	4.97	31	241.06
April	5.55	30	208.09
May	5.39	31	192.12

June	4.73	30	175.80
July	5.43	31	189.08
August	5.44	31	210.16
September	5.21	30	226.99
October	5.50	31	252.06
November	4.45	30	250.69
December	4.16	31	261.35
Total			2702.91

Sun path diagrams help in knowing how the sun will impact a selected site and building throughout the year. Stereographic sun path diagrams can be used to read the solar azimuth and altitude for a given location. Sun-path diagrams map the path of the sun across the sky. They show the position of the sun relative to the site, both by time of day and time of year. The Sun-path diagram for a twelve month period for Miri has been determined and presented as in figure 2.

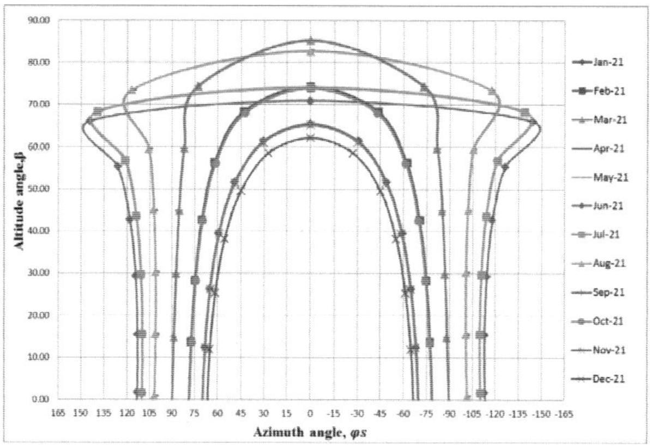

Figure 2. Sun-path diagram for Miri, Sarawak.

3. Proposed Solar Power Plant

A total solar power plant capacity of 208 kW for four buildings, which is equivalent to 52 kW per building, is being proposed to accommodate a minimum of 30% of the total electrical consumption of Curtin. The identified four buildings are Prinia 1, Prinia 2, Prinia 3 and Heron 1.

The assumptions made for the solar panel modules are presented in table 2.

Table 2. Assumptions Made.

Cost per 250 W solar panel	RM 1200
Cost per 12 kW inverter	RM 10,000
Maintenance cost per year	RM 4,000
Additional cost and labour cost per kW	RM 1,000
Life Expectancy	25

Technical Data:

Solar Module

Nice Sun PV Co., Ltd.; Model: NCS-6BM250

26 string x 8 modules

Voc = 37.5 V ; Vmp = 30.1 Volts
Isc = 8.87 A ; Imp = 8.32 A

Inverter

GOODWE Inverter GW12k-DT
5 units AC SPD 20kA
Pdcmax : 12300 W Pacmax : 12000 W
Vdcmax : 1000 V Idcmax : 22/11 A
Iacmax : 19 A 2 MPPT

Figure 3 shows the electrical interconnections of a 52 kW solar system of Prinia 1. The energy converted from solar to electricity per month by the 208 kW solar power plants has been determined and presented in table 3.

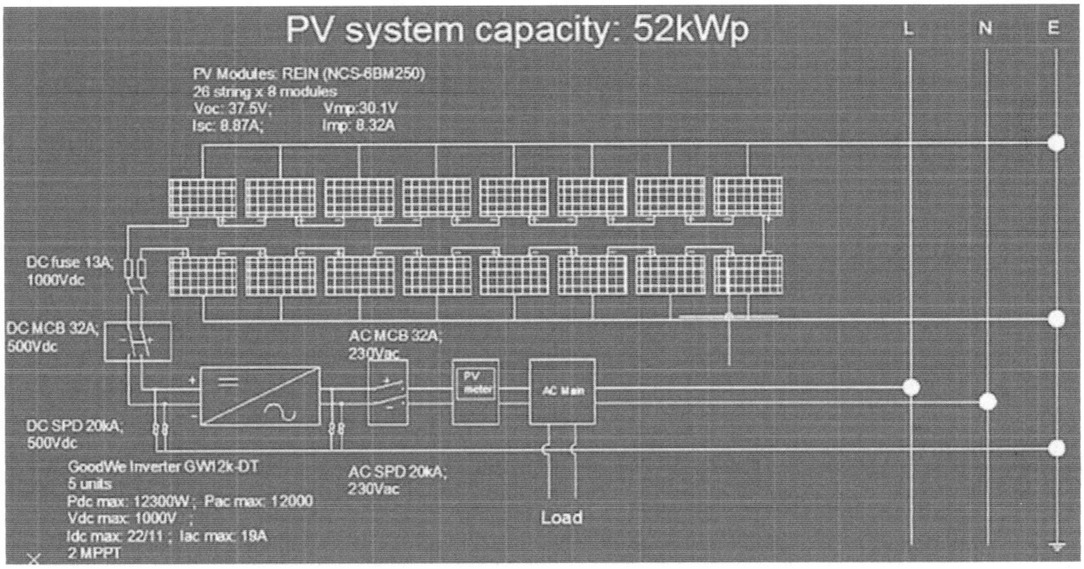

Figure 3. Prinia 1 – 52 kW Solar Modules Interconnections.

Table 3. Monthly Energy Generation (kWh).

Month	kWh
January	32111
February	26557
March	32047
April	34632
May	34755
June	29515
July	35013
August	35077
September	32510
October	35464
November	27768

December	26824
Total	382273

Though Malaysia is a tropical country with four seasons, there is not wide variation in the monthly kWh generations. The annual kWh generation is 382273 kWh and the average daily kWh generation is found to be 1047 kWh. The daily kWh generation is not very high since the modules tilting angle follows the roof angle of the buildings as it is.

4. Curtin's Energy Consumption

Curtin's monthly energy consumption for year 2014 has been received [21] and adjusted by the energy generations of the proposed 208 kW solar power system. The net energy consumption payable to the external grid is found out and tabulated as in table 4. Figure 4 shows the energy consumption with and without the PV panels.

Table 4. Net Energy Consumption Payable.

Months	Energy Consumption, kWh	Energy Generated, kWh	Net Energy Consumption Payable, kWh
January	67549	32111	35438
February	63610	26557	37053
March	76418	32047	44371
April	84539	34632	49907
May	87539	34755	52784
June	85553	29515	56038
July	75164	35013	40151
August	88505	35077	53428
September	82790	32510	50280
October	83047	35464	47583
November	87763	27768	59995
December	63930	26824	37106

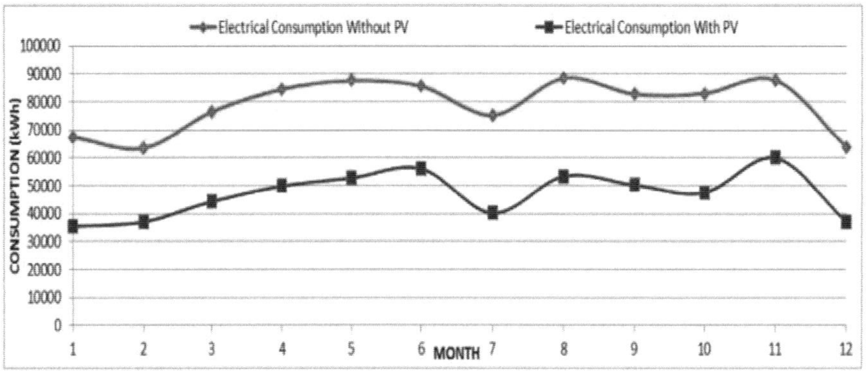

Figure 4. Energy consumption with and without solar power.

Likewise the energy consumption bills for the twelve month period with and without the proposed solar system have been worked out and the saving in energy consumption due to the solar system if implemented has also been brought out as in table 5. The comparison of electrical consumption charges with and without the proposed solar system is shown in figure 5.

Table 5. Energy Bill Comparison & Net Saving.

Months	Energy Bill Without Solar Energy, RM	Energy Bill With Solar Energy, RM	Net Saving in Energy Consumption Charges, RM
January	20470.70	10837.39	9633.31
February	19289.00	11321.77	7967.23
March	23131.40	13517.43	9613.97
April	25567.70	15178.10	10389.60
May	26467.70	16041.28	10426.42
June	25871.90	17017.34	8854.56
July	22755.20	12251.42	10503.78
August	26757.50	16234.36	10523.14
September	25043.00	15289.88	9753.12
October	25120.10	14480.90	10639.20
November	26534.90	18204.50	8330.40
December	19385.00	11337.90	8047.10
Total	286394.10	171712.27	114681.83

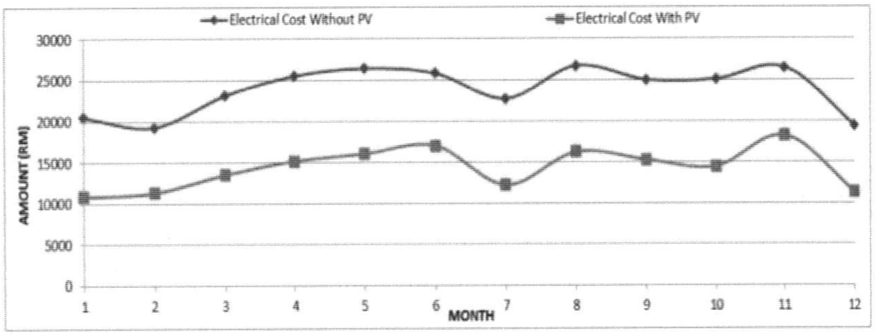

Figure 5. Energy consumption charges with and without solar power.

5. Financial and Environmental Facets

The total installation cost involved in the proposed 208 kW solar project is worked out based on the assumptions presented in table 2 as:

$$Installation\ Cost = Cost\ of\ solar\ panels + Cost\ of\ Inverters + Additional\ cost\ and\ labor$$
$$= \left[\left(\frac{208 \times 1000}{250}\right) \times 1200\right] + \left[\frac{208}{12} \times 10000\right] + [208 \times 1000]$$

Table 6 shows the total cost of the proposed solar project in terms of installation cost and maintenance cost.

Table 6. Installation Cost.

Installation Cost (RM)	1,379,733
Estimated Maintenance Costs (RM)	100,000
Total Cost (RM)	1,479,733

Assuming a life-time of 25 years, and the drop in PV panel performance of 0.8% per annum, the financial benefits out of the proposed project are determined and shown in table 7.

During the project's life span, the avoided annual carbon dioxide emission is projected as shown in table 8 [23, 24]. The reduction in carbon emission at the beginning of the project is around 286000 kg per year and increases to a maximum of around 367000 kg at the end of its life span. If there is any emission credit facility available by the norms of the local government, then this project is eligible for such a privilege.

Table 7. Financial Benefits over THE LIFE-Time.

Year	Expected PV Performance (%)	Financial Benefits (RM)
1	100.0	114681.84
2	99.2	228446.23
3	98.4	341293.16
4	97.6	453222.63
5	96.8	564234.65
6	96.0	674329.22
7	95.2	783506.33
8	94.4	891765.99
9	93.6	999108.19
10	92.8	1105532.94
11	92.0	1211040.23
12	91.2	1315630.07
13	90.4	1419302.45
14	89.6	1522057.38
15	88.8	1623894.85
16	88.0	1724814.87
17	87.2	1824817.44
18	86.4	1923902.55
19	85.6	2022070.20
20	84.8	2119320.40
21	84.0	2215653.15
22	83.2	2311068.44
23	82.4	2405566.28
24	81.6	2499146.66
25	80.8	2591809.58

Table 8. Avoided Emission.

Year	Carbon Emission Without PV (kg)	Carbon Emission With PV (kg)	Carbon Reduction (kg)
1	708670	422424	286246
2	708670	419044	289625
3	708670	415665	293005
4	708670	412286	296384

5	708670	408906	299763
6	708670	405527	303143
7	708670	402147	306522
8	708670	398768	309902
9	708670	395389	313281
10	708670	392009	316660
11	708670	388630	320040
12	708670	385250	323419
13	708670	381871	326799
14	708670	378492	330178
15	708670	375112	333557
16	708670	371733	336937
17	708670	368353	340316
18	708670	364974	343695
19	708670	361595	347075
20	708670	358215	350454
21	708670	354836	353834
22	708670	351457	357213
23	708670	348077	360592
24	708670	344698	363972
25	708670	341318	367351
Total	17716739	9546775	8169964

6. Conclusions

The paper has addressed a renewable energy concept to reduce the fossil fuel energy conception of Curtin University, Sarawak, Malaysia. The energy generated by the 208 kW PV project varies from a minimum of 31% to a maximum of 47% of the monthly energy consumption of Curtin. Hence there is corresponding reduction in the energy bill payable to the grid connected electricity. Apart from the financial aspect, there is an added advantage of emission credit by avoiding the carbon emission throughout the project lifespan. The proposed 208 kW solar system saves an energy consumption of more than 380,000 kWh per year, and a CO2 offset by 285 Tons per year. The only concern is the budget for the investment of the proposed project. Curtin shall implement the project in four segments considering one building at a time.

References

[1] Amanda Christy Brown and Kristin Mcginn Manhoney 2009 *The Story of Stuff* ed Braelan Murray (Los Angeles : Los Angeles Times)

[2] Margot Roosevelt 2010 *Teaching 'stuff' about ecology* (Los Angeles : Los Angeles Times)

[3] Paul Hawken, Amory Lovins and Hunter Lovins L 1999 *Natural Capitalism* (New York: Hachette Book Group Publishing)

[4] Schmidt-Bleek F 2009 *The Earth: Natural Resources and Human Intervention* ed Klaus Wiegandt (London: Haus Publishing Limited)

[5] Seri 2009 *Over Consumption? Our use of the world's natural resources* (Austria: The Sustainable Europe Research Institute) http://www.foe.co.uk/sites/default/files/downloads/overconsumpti on.pdf

[6] Betsy Taylor and Dave Tilford 2000 *Why Consumption Matters* ed Juliet B Schor and Douglas Holt (San Francisco: The Consumer Society Reader) p 467

[7] Palanichamy C, Sundar Babu N, and Nadarajan C 1999 Privatizing and restructuring Indian power sector: an overview *Journal of Institution of Engineers (India. Electrical Engineering)*

80 23–30
[8] Aigner T, Jaehnert S, Doorman G L and Gjengedal T 2012 The effect of large-scale wind power on system balancing in Northern Europe *IEEE Trans. on Sustainable Energy* **3(4)** 751–59
[9] Palanichamy C, Babu N S and Nadarajan C 2004 Renewable energy investment opportunities in Mauritius-an investor's perspective *Renewable energy* **29(5)** 703–16
[10] Dolf Gielen 2012 *Renewable energy technologies: cost analysis series* vol 1 ed Zuzana Dobrotkova (Germany:International Renewable Energy Agency) Power Sector Issue 2/5
[11] Hayward J Paul W Graham and Peter K Campbell 2011 *Projections of the future costs of electricity generation technologies: an application of CSIRO's global and local learning model (GALLM)* (Australia: CSIRO)
[12] Kolb GJ, Ho C K, Mancini T R and Gary J A 2011 *Power Tower Technology Roadmap and Cost Reduction Plan* (Alburqurque: New Mexico/ Sandia National Laboratories)
[13] Palanichamy C, and N S Babu 2005 A visual package for educating preparatory transmission line series compensation," *IEEE Transactions on Education*, **48(1)** 16-22
[14] ATKearney and ESTELA 2010 *Solar Thermal Electricity 2025* (Brussels: ESTELA) http://www. estelasolar.eu/index.php?id=22
[15] Emerging Energy Research 2010 *Global Concentrated Solar Power Markets and Strategies: 2010-2025* (Cambridge: MA/ IHS Emerging Energy Research)
[16] NREL 2012 *Concentrating Solar Power Projects Database* (New York: US Department of Energy) Available at: http:// www.nrel.gov/csp/solarpaces/by_country.cfm.
[17] Muller-Steinhagen, H. 2011 *Solar Thermal Power Plants-On the way to commercial market introduction* (Stuttgart: German Aerospace Centre (DLR)
[18] Mekhilef S, Safari A, Mustaffa W E S, Saidur R, Omar R and Younis M A A 2012 Solar energy in Malaysia *Renewable and Sustainable Energy Reviews* **16** 386–96
[19] Song Eng Eng and EXIM Bank *IA/Solar Panel Industry in Malaysia (Malaysia:* Song Eng Eng*)* Available at: http://www.exim.com.my/sites/default/files/industry_assessment_-_solar_panel_industry_in_malaysia_january_2015.pdf
[20] GEMAS 2014 *Amcorp commissions largest single site solar power plant in Malaysia* in (Malaysia: Borneo Post)
[21] Curtin University year no Curtin University Sarawak (Malaysia: Sarawak/Curtin University) Available at: www.curtin.edu.my
[22] Miri Council 2015 Miri Council in Brief. Official Website of Miri City Council (Malaysia: Miri Council) Available at http://www.miricouncil.gov.my/modules/web/ Retrieved 3 Sept. 2015
[23] Palanichamy C, Babu N S and Nadarajan C 1991 Simple algorithm for economic power dispatch *Electric Power Systems Research* **21(2)** 147–53
[24] Naveed Ahmed Khan N A, Awan A B, Mahmood A, Razzaq S, Zafar A and Sidhu G A S 2015 Combined emission economic dispatch of power system including solar photo voltaic generation *Energy Conversion and Management* **92(1)** 82–91

IOP Conf. Series: Materials Science and Engineering **121** (2016) 012012 doi:10.1088/1757-899X/121/1/012012

A solar power plant for Curtin University Malaysia

A/Prof Dr C Palanichamy

Electrical and Computer Engineering Department, Curtin University, Sarawak, Miri, Malaysia

E-mail: drcpc@curtin.edu.my

Abstract. The Curtin University, Sarawak Malaysia (Curtin Sarawak) is the first and largest offshore campus of Curtin University in Perth, Western Australia, and the first foreign university to be established in East Malaysia in partnership with the Sarawak State Government. Today's major concern of Curtin is its monthly electrical energy consumption and the electricity bill since its monthly energy consumption exceeds 0.3 Million kWh, and the corresponding electricity bill surpasses RM 95000. Such a situation necessitates Curtin to curtail the heavy energy consumption with immediate effect. Introducing Renewable Energy Source such as PV Solar Systems is a cost-effective and environmental friendly solution to reduce the exponential increase in energy consumption charges of Curtin. Hence, this paper proposes a 90 kW solar power plant for Curtin Sarawak.

1. Introduction

In recent years, copious studies and researches [1-6] highlight on renewable energy sources such as solar, hydro, wind and biomass. The primary objective on these researches is to upsurge the feasibility of renewable energy in substituting the conformist manner of power generation through burning of fossil fuels. The reason behind is not only due to environment pollution, but also the depletion of fossil fuels, its ever-increasing demands, and time-varying prices.

Among innumerable renewable energy technologies, solar power generation [4-6] is widely-accepted due to its lower costs, higher efficiency, and smaller occupation of space. In Asia and Australia, solar power generation [2,5,7,8] is dominant among other types of renewable energies. This is due to the promising availability of sunlight in the region, which can readily be harnessed and converted into electrical power for consumption.

Among all of the renewable energy sources, solar energy acts as an ideal alternative energy source because it can generate electricity directly from the sun. The main advantages of using solar energy are the environmental benefits, higher efficiency, high reliability due to non-rotating conversion principle, and higher life span of solar panels. Besides that, it can be set on or off the grid.

Malaysia is moving towards generating electricity using renewable energy. In order to promote the growth of renewable energy sector in Malaysia, Feed-in Tariff (FiT) mechanism [9,10] has been introduced by Malaysian government in 2011, in accordance with Renewable Energy Act 2011 and Sustainable Energy Development Authority Act 2011. A Feed-in Tariff (FiT) is a policy mechanism designed to promote investment in Renewable Energy technologies by offering long-term power purchasing contracts to renewable energy producers at a fixed premium rate. The FiT is financed by the Malaysian electricity consumers who contribute 1% of their total electricity bill towards a Renewable Energy Fund when they use more than 300 kWh of electricity per month. The approved renewable energy power producers will enter into a Renewable Energy Power Purchase Agreement with the Distribution Licensee (Tenaga Nasional Berhad) and will be paid for every kilowatt hour of power exported to the grid for 21 years.

The combination of the fall in the cost of solar panels and attractive tariffs has made investments in solar energy generation lucrative in Malaysia [11-13]. As a result, solar energy in operation grew by an average of 124% per annum since the inception of FiT. Under the 10th Malaysia Plan, Malaysia plans to achieve 985 MW or 5.5% share of renewable energy in the energy mix by 2015. By 2020, the target is for renewable energy to comprise 11% or 2,080 MW of overall electricity generation in the country. However, as to-date, renewable energy has yet to be considered a major source of electricity generation as it contributes less than 1% to the energy mix in Malaysia.

Following Malaysia's renewable energy plans, Curtin University, Sarawak Malaysia (Curtin Sarawak) [14-16], the first and largest offshore campus of Curtin University in Perth, Western Australia, and the first foreign university to be established in East Malaysia in partnership with the Sarawak State Government has shown keen interest in introducing renewable energy systems. As a support to Curtin's interest, this paper proposes a 90 kW solar power plant.

2. Curtin Sarawak Data

2.1. Sun-path Diagram

Sun-path diagrams map the path of the sun across the sky. They show the position of the sun relative to the site, both by time of day and time of year. The position of the sun in the sky relative to an observer on Earth is defined by its altitude angle α (solar elevation angle) and its azimuth angle Ψ [17]. Both the angles with respect to an observer are shown in figure 1.

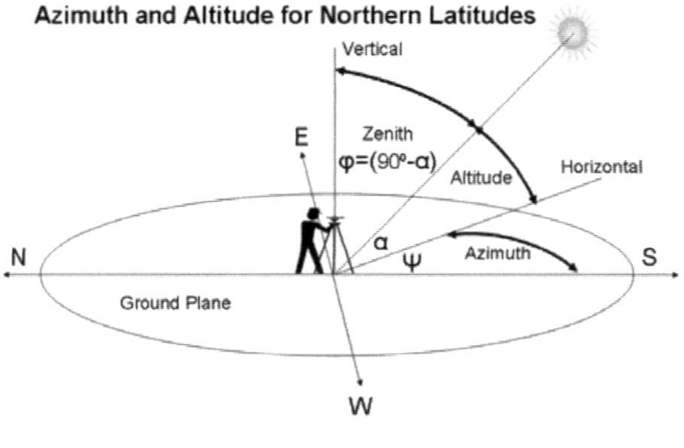

Figure 1. Azimuth and Altitude Angles.

For Miri, Sarawak, the relevant information to determine the Sun-path diagram is as shown:

Latitude: +4.4 (4°24'00"N)

Longitude: +113.97 (113°58'12"E)

Time zone: UTC+8 hours

Country: Malaysia

Continent: Asia

Sub-region: South-Eastern Asia

The Sun-path diagram for a twelve month period for Miri has been determined and presented as in figure 2.

2.2. Curtin Energy Consumption

For the twelve month period in year 2013, the monthly kWh, kVAR consumptions and the corresponding energy consumption charges are depicted in table 1. The minimum energy consumption charges are noticed during January and February months, which are semester break months. The maximum energy consumption of 100530 kWh occurs during September month, which is a semester on period, and the corresponding electricity charge is found to be RM 126217.95. It is observed that during the semester break period, practically the campus is with few students; still the energy consumption charges are almost 50% of the semester on period. The daily energy consumption is shown in table 2.

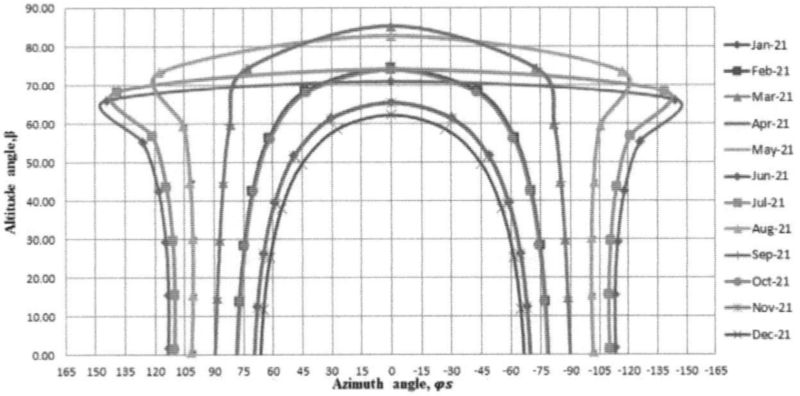

Figure 2. Sun-path diagram for Miri, Sarawak.

Table 1. Curtin Energy Consumption and Charges.

Month	kWh	kVAR	Charges(RM)
January	204566	49003	62609.50
February	200899	51339	61531.00
March	359248	89203	109521.00
April	307884	75065	94073.45
May	338445	86274	103479.45
June	318878	83927	97033.55
July	307304	78938	93965.20
August	335515	85370	102548.30
September	413847	100530	126217.95
October	288578	71784	88353.40
November	321711	80731	98434.10
December	267824	69746	81877.90
Total	3664699	921910	1119645.00

Table 2. Daily Energy Consumption and Charges.

Month	kWh	kVAR	Charges(RM)
January	6818.87	1633.43	2086.98
February	6696.63	1711.30	2051.03
March	11974.93	2973.43	3650.70
April	10262.80	2502.17	3135.78

May	11281.50	2875.80	3449.32
June	10629.27	2797.57	3234.45
July	10243.47	2631.27	3132.17
August	11183.83	2845.67	3418.28
September	13794.90	3351.00	4207.27
October	9619.27	2392.80	2945.11
November	10723.70	2691.03	3281.14
December	8927.47	2324.87	2729.26

3. Proposed Solar Power Plant

3.1. Proposed Site

Curtin campus map is shown in figure 3. Figure 4 shows the newly built Chancellery building. This is the ideal site chosen for the proposed solar power plant since no tall raised buildings and trees nearby the building; hence no obstacle for the sunlight incidence on the roof-top. Also there won't be any shading effect reducing the energy capture.

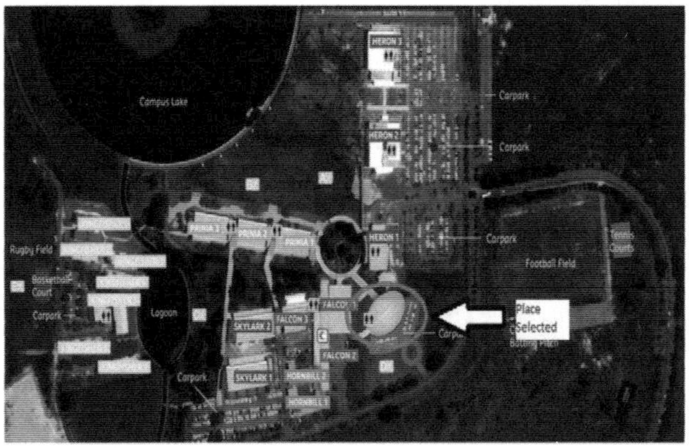

Figure 3. Curtin Campus Map.

Figure 4. Chancellery Building – Proposed Site.

Figure 5 shows the roof-top area of the Chancellery building. This is the place identified for the proposed 90 kW solar power plant.

3.2. Proposed Solar Panel

The module chosen for this project is Yingli Solar YL255P-29b (255Wp) Polycrystalline with technical information as shown in figure 6. This is due to Yingli Solar (NYSE: YGE) [18] is the world's largest solar panel manufacturer. More than 50 million Yingli solar panels (representing over 13 gigawatts) have been shipped to more than 90 countries. Yingli covers the entire solar panel production process, from polysilicon to ingots and from solar cells to quality solar panels, and its manufacturing capacity has reached to 2.45GW today. This solar power panel has high efficiency, multi-crystalline silicon solar cells with high transmission and textured glass deliver a module efficiency up to 16.2%. This will minimize installation costs and maximize the kWh output of the system per unit area. This kind of solar panel withstands wind loads of up to 2.4 kPa and snow loads of up to 5.4 kPa, confirming mechanical stability. Besides that, it also successfully endures ammonia and salt-mist exposure at the highest severity level, ensuring their performance in adverse conditions. This solar power panel will have 10-year limited product warranty as well as limited power warranty such as 10 years at 91.2% of the minimal rated power output and 25 years at 80.7% of the minimal rated power output.

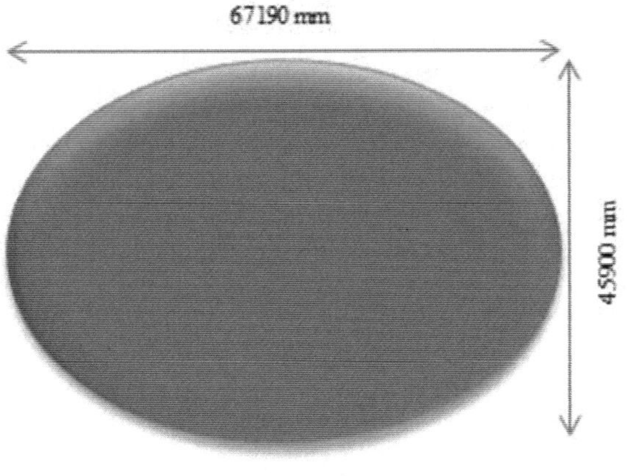

Figure 5. Chancellery Building – Roof-top.

PV Panel	Yingli YL255P-29b Poly (60 cells)
Pmax	255Wp (rated power output)
Isc	9.01A (short-circuit current)
Voc	37.7V (open-circuit voltage)
Impp	8.49A (current at Pmax)
Vmpp	30.0V (voltage at Pmax)
Modules Efficiency	15.6%
Maximum System Voltage	1000V
Power Output Tolerance	0/+4.99W (only positive tolerance)
Background Coloration	White
Frame Color	Silver
Connections	MC4 / IP67 (110cm cables / 4mm)
Load Capacity	Pa max. 2400 N/m2 for winds and 5400 N/m2 for snow
Dimensions	1650x990x40mm
Weight	18.5 kg
Product Warranty	10 years
Performance Warranty	10 years for 92% and 25 years for 82% of rated power

Figure 6. Module – Technical Information.

Figure 7 presents the technical information of the selected module and the physical module dimension is 1650mmx990 mm.

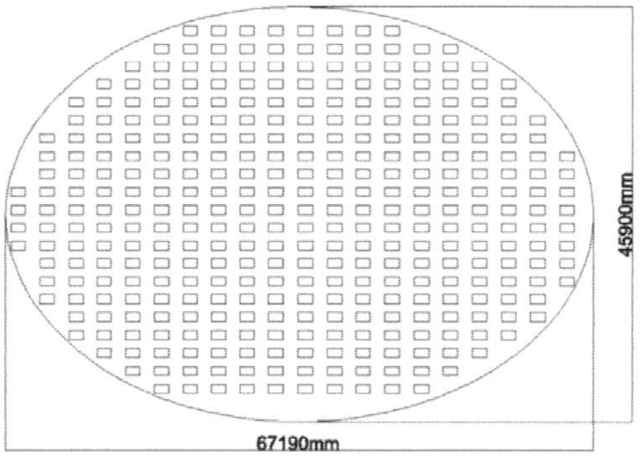

Figure 7. Proposal One – Roof with 341 Modules.

3.3. Proposed Inverter

The SMA STP20000TL-30 is the ideal inverter for large-scale and commercial and industrial plants [19,20]. It contains cutting edge grid management functions such as Integrated Plant Control. This feature allows the inverter to adjust reactive power at the point of normal coupling and removes the need for a discrete reactive power controller. In addition, it is capable of supporting multistring configurations and provides flexibility in the system design. Some advantages of the SMA STP20000TL-30 inverter include: (i) 98.4% maximum efficiency, (ii) DC input voltage tolerance of 1000V, (iii) Supports SPD type II DC surge arrestor for better safety, and (iv) Supports SMA Sunny Portal and multi-functional relay.

4. Proposal

4.1. Modules on Roof-top

All the solar panels are placed horizontally following the Chancellery building's roof architecture. Since all are placed horizontally, the tilting angle of the module is 0°. Considering the area of each module and the area of the Chancellery roof, 341 numbers of Yingli Solar YL255P-29b (255Wp) Polycrystalline modules are needed. Plan view of design one is shown in figure 8.

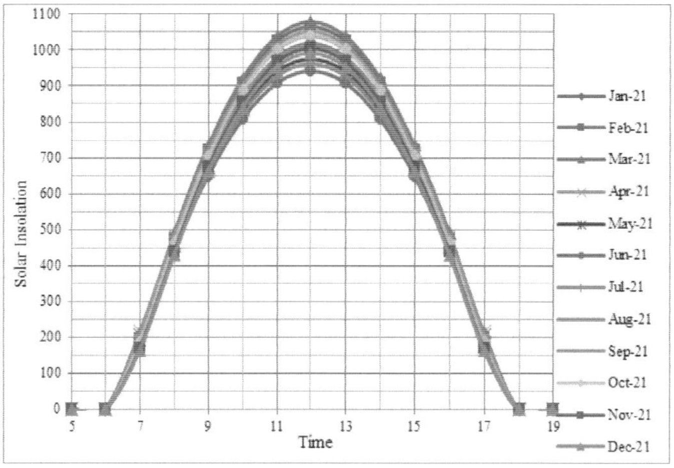

Figure 8. Monthly Insolation on 21st Day of Every Month.

4.2. Curtin's Solar Insolation

The solar insolation on the roof-top modules with 0° module tilt angle has been determined on 21st day of every month and shown in figure 9. Besides the daily and monthly insolation on unit area of the solar modules are determined and presented in table 3.

Table 3. Insolation on the Roof-top Modules.

Month	Days/ Month	Daily Isolation	Monthly Isolation
January	31	7318.51	226.87
February	28	7719.33	216.14
March	31	7870.15	243.97
April	30	7627.63	228.83
May	31	7198.24	223.15
June	30	6947.79	208.43
July	31	7078.34	219.43
August	31	7442.38	230.71
September	30	7669.17	233.08
October	31	7542.05	231.01
November	30	7217.29	216.52
December	31	7082.26	219.55
Total			2697.49

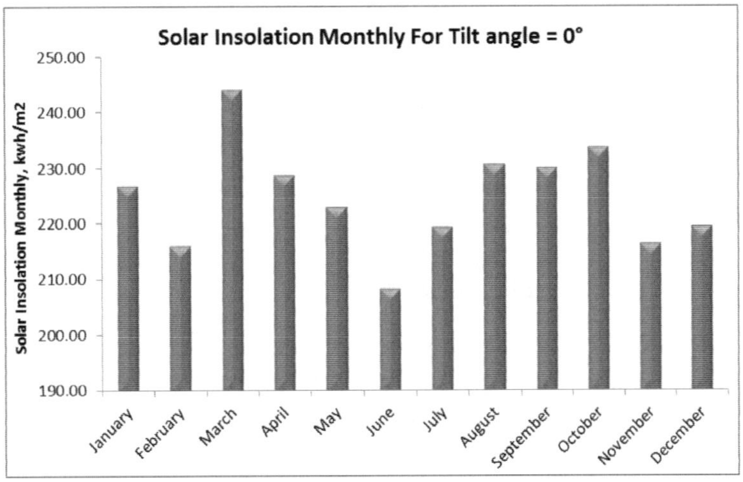

Figure 9. Monthly Insolation on the Proposed Roof-top.

4.3. Annual Energy Generated

Considering the net solar insolation of 90%, the total number of solar modules, the efficiencies of the module and the inverter, the annual energy exportable to the grid has been determined and presented in table 4.

Table 4. Insolation on the Roof-top Modules.

Solar insolation per year (kWh/m2)	2697.49	2697.49
Net insolation per year (90%) (kWh/m2)	2427.74	2427.74

Total number of PV module	324	17
Area of one PV module (m2)	1.634	1.634
Total Area of the PV array (m2)	529.416	27.778
Annual solar energy (kWh)	1285284	67438
Efficiency of solar power panel (%)	15.6	15.6
Efficiency of the inverter (%)	98.5	97.0
Annual Solar energy exportable to the grid (kWh)	197497	10205
Total energy exportable to the grid by all the 341 modules per annum (kWh)		207702

4.4. Financial Aspects
The capital cost of the proposed 90 kW solar power plant has been estimated and shown in table 5.

Table 5. Financial aspects.

Item	No.	Price per unit	Total Price
Yingli Solar YL255P-29b (255Wp) Poly	341	RM 986	RM 336185
SMA Sunny Tri Power 20 kW Inverter STP20000TL-30	4	RM 22313	RM 89250
SMA Sunny Boy 4 kW Inverter SB4000TL-21	1	RM 8509	RM 8509
Total equipment cost			RM 433944
Installation cost			RM 73256
Total project cost			RM 507200

The expected generation of the power plant per year is 207702 kWh. The Feed-in Tariff of sustainable energy development authority (SEDA) is RM 0.3504/kWh. The generated revenues and profit are calculated as:

Total revenues generated = 207702 kWh x RM 0.3504/ kWh = RM 72778.78 per year. Considering 12.5% as maintenance and depreciation charges, the net revenue per year = RM 63681. Hence the payback period = 7.96 say 8 years. In terms of Curtin's annual energy consumption, the proposed 90 kW solar power plant reduces the energy consumption by 207702 kWh or 5.67%.

5. Conclusions
This paper has proposed a solar power project of capacity 90 kW following the renewable energy target of Malaysian Government. The newly constructed Chancellery building has been chosen as the site for consideration since no obstacles are noticed around the building. As Malaysia is a tropical country, the solar insolation thorough out a year is good, and the energy produced by solar panels is found to be attractive even with solar modules of 15% efficiency. The success of a solar power project mainly relies on the site selection, the module & inverter selection and the energy purchase tariff. As far as the proposed project is concerned, the solar energy converted into electrical energy is found to be more than 2300 kWh per kW capacity installed. Moreover, the payback period is highly encouraging as 8 years with an internal rate of return as 12.5%.

References
[1] GEMAS 9 June 92014 *Amcorp solar power plant largest in Malaysia* in The Rakyat Post Malaysia
[2] Palanichamy C and Srikrishna K 1991 Simple algorithm for economic power dispatch *Electric Power Systems Research* **21(2)** 147–53

[3] Lai N Y G, Yap E H and Lee C W 2011 Viability of CCS: a broad-based assessment for Malaysia *Renewable and Sustainable Energy Reviews* **15(8)** 3608–16

[4] Ikram M, Imran M, Nunzi J M, Bobbara S R, Salamat Ali1 and Islah-u-din 2015 Efficient and low cost inverted hybrid bulk heterojunction solar cells *Journal of Renewable Sustainable Energy* ***7(4)*** 043148(2015)

[5] ARENA 2013 *Australian Renewable Energy Agency. About* (Australia: *Australian Renewable Energy Agency)* Available at: www.arena.gov.au/about-arena

[6] Aigner T, Jaehnert S, Doorman G L and Gjengedal T 2012 The effect of large-scale wind power on system balancing in Northern Europe *IEEE Trans. on Sustainable Energy* **3(4)**751–59

[7] Gomesh N, Daut I, Irwanto M, Irwan Y M and Fitra M 2013 Study on Malaysian's perspective towards renewable energy mainly on solar energy *Energy Procedia* **36** 303–12

[8] Palanichamy C, Meseret Nasir and Veeramani 2014 Wind Cannot be Directed But Sails Can be Adjusted for Malaysian Renewable Energy Progress *IOP Conf. Series: Materials Science and Engineering (Miri)* Vol 78 (Miri: Sarawak/ IOP Publishing) 012028

[9] Wonga S L, Norzita Ngadia, Tuan Amran Tuan Abdullahb, Inuwac I M 2015 Recent advances of feed-in tariff in Malaysia *Renewable and Sustainable Energy Reviews* **41** 42–52

[10] mySolarPower year no *Feed-in Tariff (FiT)* (Malaysia: mySolarPower Sdn Bhd) http://www.sol arpower.my/feed-in-tariff/

[11] Mekhilef S, Safari A, Mustaffa W E S, Saidur R, Omar R and Younis M A A 2012 Solar energy in Malaysia *Renewable and Sustainable Energy Reviews* **16** 386–96

[12] Song Eng Eng and EXIM Bank *IA/Solar Panel Industry in Malaysia (Malaysia:* Song Eng Eng*)* Available at: http://www.exim.com.my/sites/default/files/industry_assessment_-_solar_panel _industry_in_malaysia_january_2015.pdf

[13] GEMAS 2014 *Amcorp commissions largest single site solar power plant in Malaysia* in (Malaysia: Borneo Post)

[14] Curtin University 2014 Curtin University Sarawak (Malaysia: Sarawak/Curtin University) Available at: www.curtin.edu.my

[15] Palanichamy C and Babu N S 2005 A visual package for educating preparatory transmission line series compensation *IEEE Transactions on Education* **48(1)** 16–22

[16] Tang Fu E 2012 An energy consumption study for a Malaysian University *International Journal of Environmental, Chemical, Ecological, Geological and Geophysical Engineering* **6(8)**

[17] Electropedia year no *Solar Power (Technology and Economics)* (UK: Wood bank Communications Ltd 2005) Available at: http://www.mpoweruk.com/solar_power.htm

[18] Yingli Solar 2014 *Yingli Solar (NYSE: YGE)* (China: Yingli Green Energy Holding Co., Ltd) Available at: http://www.yinglisolar.com/as/about/

[19] SMA Solar Technology AG 2014 *Sunny Design* (Germany: Niestetal/SMA Solar Technology AG*)* Available at: http://www.sunnydesignweb.com/sdweb/#/Home

[20] Sol Distribution 2014 *Sunny-Tripower-Inverters* (Australia: Sol Distribution Pty Ltd) Available at: http://www.sol-distribution.com.au/products/Sunny-Tripower-Inverters/sma-stp20000tl-3 0-inverter.html

Hybrid estimation technique for predicting butene concentration in polyethylene reactor

Jarinah Mohd Ali[1] and M.A. Hussain[1]

[1]Department of Chemical Engineering

Faculty of Engineering University of Malaya

50603, Kuala Lumpur,

Malaysia.

jarinah@siswa.um.edu.my

Abstract. A component of artificial intelligence (AI), which is fuzzy logic, is combined with the so-called conventional sliding mode observer (SMO) to establish a hybrid type estimator to predict the butene concentration in the polyethylene production reactor. Butene or co-monomer concentration is another significant parameter in the polymerization process since it will affect the molecular weight distribution of the polymer produced. The hybrid estimator offers straightforward formulation of SMO and its combination with the fuzzy logic rules. The error resulted from the SMO estimation will be manipulated using the fuzzy rules to enhance the performance, thus improved on the convergence rate. This hybrid estimation is able to estimate the butene concentration satisfactorily despite the present of noise in the process.

1. Introduction

Butene or co-monomer concentration is also another favorable parameter besides monomer concentration in polymerization process since it tends to affect the molecular weight distribution (MWD) of the polymer produced. The lower the distribution of the co-monomer the higher the MWD of the polymer [1]. Therefore, it is also crucial to monitor the distribution of this co-monomer. So far, there is no effort in estimating this parameter in the polymerization reactor since many researchers are focusing on the estimation of monomer and other parameters including reaction rate and chain length, which will give higher impact to the process. Concentrations are estimated using estimators prior to implement the control procedure in a polymerization reactor [2] [3, 4] [5] [6] [7].

Estimator will estimate the unknown variables better than the so-called expensive sensors especially in terms of accuracy and estimation time [8]. In certain cases, the conventional estimator will be merged with other estimator to avoid offsets and improved the convergence rate. The combination can be either between two or more conventional estimators or with artificial intelligence (AI). AI has also been applied by several researchers as estimators and in some cases they are better than the conventional estimator and are easier to be formulated [9].

For this work, we combined the conventional sliding mode observer (SMO) with fuzzy logic to estimate the co-monomer concentration in an ethylene polymerization reactor. SMO is and extended version of extended Luenberger observer that have been proven to provide fast estimation and is also a stable estimator without require any input assumption during the estimation process [10]. On the other hand, fuzzy logic is the simplest algorithm of AI, suitable to be combined with those conventional estimators by manipulating only the IF and THEN rules for enhancing the estimation performances and overcoming the limitations of the single SMO. Fuzzy rules are applied based on the estimation error, and the change of error to obtain desired results faster with minimal efforts.

Introduction is given in this section follow by the ethylene polymerization reactor model in section 2. The hybrid estimator design according to a case study is discussed in section 3 while conclusion is in section 4.

2. Ethylene polymerization reactor model

The ethylene polymerization reactor model for this work is based on the well-mixed UNIPOL model developed by McAuley (1990) [11, 12]. The reactor is illustrated in figure 1 where the feed gas is combined with the recycled gas before entering at the bottom of the fluidized bed reactor. Another input stream is the feed gas containing the Ziegler-Natta catalyst. Butene is one of the four major components entering the reactor. The other components are the monomer (ethylene), hydrogen (H2) and nitrogen (N2). The reactor temperature is controlled by manipulating both the cooling water and the feed temperature.

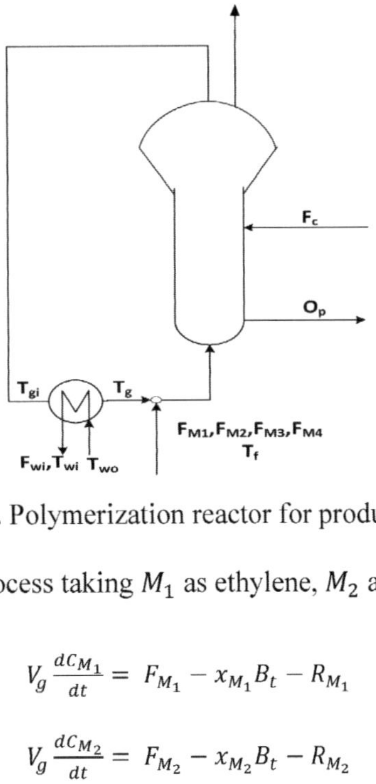

Figure 1. Polymerization reactor for producing polyethylene.

The mole balances of the process taking M_1 as ethylene, M_2 as butene, M_3 as hydrogen and M_4 as nitrogen are as follow:

$$V_g \frac{dc_{M_1}}{dt} = F_{M_1} - x_{M_1} B_t - R_{M_1} \quad (1)$$

$$V_g \frac{dc_{M_2}}{dt} = F_{M_2} - x_{M_2} B_t - R_{M_2} \quad (2)$$

$$V_g \frac{dc_{M_3}}{dt} = F_{M_3} - x_{M_3} B_t - R \quad (3)$$

$$V_g \frac{dc_{M_4}}{dt} = F_{M_4} - x_{M_4} B_t \quad (4)$$

With $$R_{M_1} = C_{M_1} Y_c k_{p1}^{e^{\frac{E}{R}(1/T - 1/T_{ref})}} \quad (5)$$

$$R_{M_2} = C_{M_2} Y_c k_{p2}^{e^{\frac{E}{R}(1/T-1/T_{ref})}} \tag{6}$$

Number of moles at the catalyst site is given by:

$$\frac{dY_c}{dt} = F_c a_c - k_d Y_c - O_p Y_c / B_w \tag{7}$$

With outlet rate as:

$$O_p = M_{w_1} R_{M_1} + M_{w_2} R_{M_2} \tag{8}$$

The total pressure of the reactor is given by:

$$P_t = (C_{M_1} + C_{M_2} + C_{M_3} + C_{M_4})RT \tag{9}$$

$$F_w C p_w (T_{wi} - T_{wo}) = 0.5UA\left[(T_{wo} + T_{wi}) - (T_{gi} + T_g)\right] \tag{10}$$

Bed temperature and recycle stream temperature are described as follows:

$$(M_r C p_r + B_w C p_p)\frac{dT}{dt} = HF + HG - HR - HT - HP \tag{11}$$

$$M_g C p_g \frac{dT_g}{dt} = F_g C p_g (T_{gi} - T_g) + F_w C p_w (T_{wi} - T_{wo}) \tag{12}$$

Where $$HF = FM_1 Cp_{M1} + FM_2 Cp_{M2} + FM_3 Cp_{M3} + FM_4 Cp_{M4} \tag{13}$$

$$HG = F_g C p_g (T_g - T_{ref}) \tag{14}$$

$$HT = (F_g + B_t) C p_g (T - T_{ref}) \tag{15}$$

$$HP = O_p C p_p (T - T_{ref}) \tag{16}$$

$$HR = M_{w_1} R_{M_1} \Delta H_r \tag{17}$$

The above equations are applied to develop the polymerization model in SIMULINK/ MATLAB to obtain the actual value that will be compared with the value from the proposed hybrid estimator. They are also used to develop the estimator's equation that we will show in section 3.

3. Hybrid estimator design

SMO is first developed and the error of estimation (e) as well as the change of error (Δe) is recorded and being used as the input to the fuzzy logic framework for developing the hybrid fuzzy-SMO estimator. The schematic diagram is given in figure 2.

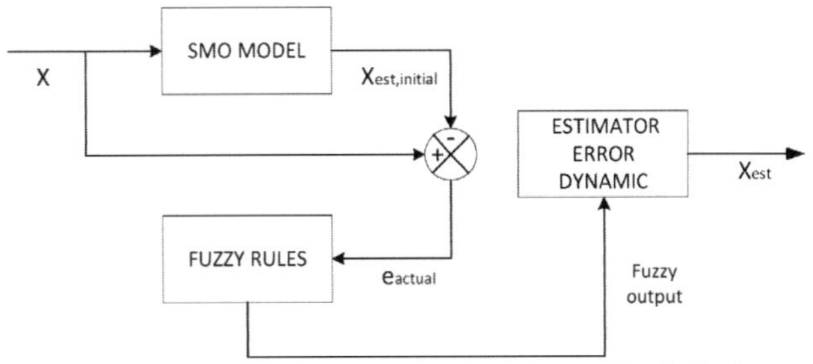

Figure 2. Schematic Diagram of Fuzzy-SMO design

In order to estimate butene concentration, the SMO equation is given below:

$$\hat{C}_{M_2} = z_2 + K_{11}sgn(T) + K_{12}sgn(T_g) \quad (18)$$

Where z_2 is as follow:

$$[z_2] = [a] - [K_{11} \quad K_{12}]\begin{bmatrix} f \\ g \end{bmatrix} \quad (19)$$

Where $$a = \frac{1}{V_g}(F_{M_2} - C_{M_2}B_t - R_{M_2}) \quad (20)$$

$$f = \frac{1}{(M_r Cp_r + B_w Cp_p)}(HF + HG - HR - HT - HP) \quad (21)$$

$$g = \frac{1}{M_g Cp_g}[F_g Cp_g(T_{gi} - T_g) + F_w Cp_w(T_{wi} - T_{wo})] \quad (22)$$

Where a is taken from the process model specifically Eq. (2) while f and g represented the measured reactor temperature (T) and recycle stream temperature (Tg) from Eq. (11) and (12) respectively.

Then the hybrid fuzzy-SMO is designed as given in Eq. (16) where e_f is the output from fuzzy logic. The input of the fuzzy logic is the error (e) and the change of error (Δe) obtained from the estimation results using the single SMO. The fuzzy logic rules are tabulated in Table 1 with e = error, Δe = change of error, NS = negative small, NL = negative large, ZO = zero, PS = positive small and PL = positive large.

$$\hat{C}_{M_2} = z_2 + K_{21}sgn(e_f) + K_{22}sgn(e_f) \quad (23)$$

Table 1. Fuzzy rules for concentration estimation

	Δe				
e	NS	NL	ZO	PS	PL
NS	PL	PL	ZO	ZO	ZO
NL	PL	PL	ZO	ZO	ZO
ZO	PS	PS	ZO	ZO	ZO
PS	ZO	ZO	NS	NL	NL
PL	ZO	ZO	NS	NL	NL

The results are given in figure 3 and figure 4 based on with and without noise conditions. Based on the figures, we observed that Fuzzy-SMO showed immediate convergence (0s) compared to SMO (20s) and able to estimate accurately despite the noise. Therefore, the proposed hybrid estimator is able to estimate the butene concentration by eliminating the offset and slow estimation time obtained using the single SMO. The hybrid Fuzzy-SMO is also in good agreement with the actual concentration during noisy condition.

This similar observer has been also applied in estimating monomer concentration and melt index with satisfactory results. It is also being compared with hybrid extended Luenberger observer (ELO) and fuzzy logic or Fuzzy-ELO [13, 14].

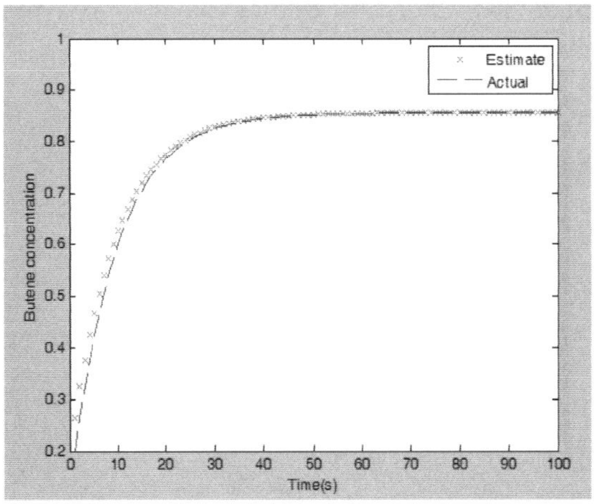

a) SMO

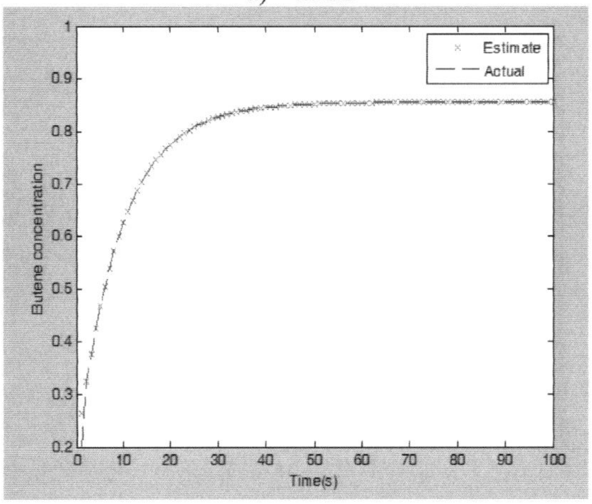

b) Fuzzy-SMO

Figure 3. Butene concentration estimation (without noise/disturbance)

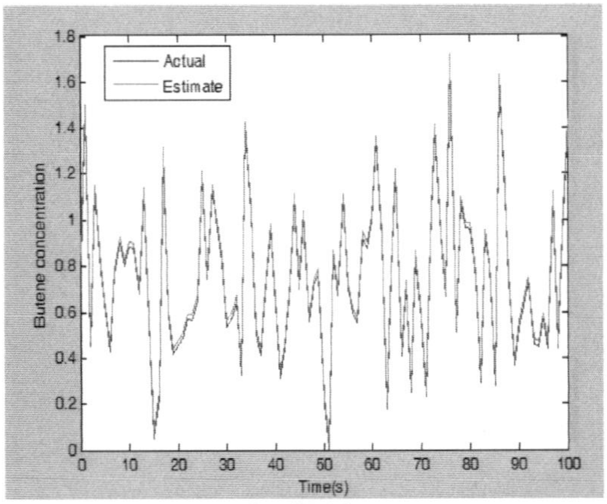

a) SMO

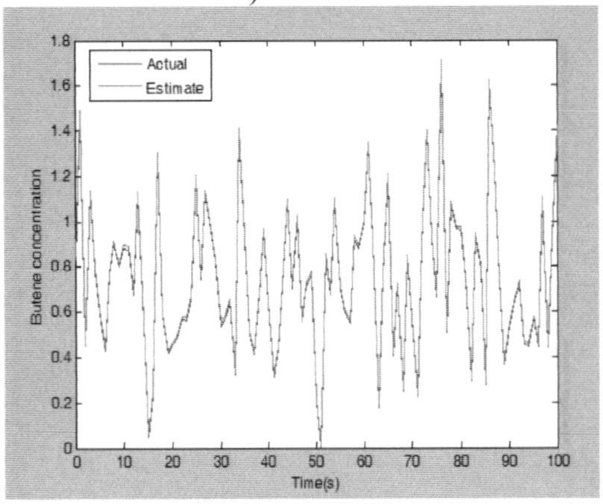

b) Fuzzy-SMO

Figure 4. Butene concentration estimation (with noise/disturbance)

4. Conclusion

Fuzzy-SMO is the best approached in predicting the butene or co monomer concentration in an ethylene polymerization reactor and able to handle noise as well as provide fast convergence compared to the single SMO. In future, the estimator will be applied to predict several other key features in a polymerization process such as melt index, heat transfer coefficient, molecular weight distribution (MWD) and reaction rate using the hybrid estimator

Acknowledgement

The authors are grateful to the University of Malaya and the Ministry of Higher Education in Malaysia for supporting this collaborative work under the high impact research grant UM.C/HIR/MOHE/ENG/25.

IOP Conf. Series: Materials Science and Engineering **121** (2016) 012013 doi:10.1088/1757-899X/121/1/012013

Nomenclature

B_t	Bleed flow rate
B_w	Polymer mass in bed
C_{M_2}	Co-monomer concentration
C_{M_1}	Monomer concentration
C_{M_4}	Nitrogen concentration
C_{M_3}	Hydrogen concentration
Cp_{M2}	Co-monomer heat capacity
Cp_{M3}	Hydrogen heat capacity
Cp_{M4}	Nitrogen heat capacity
Cp_g	Recycle gas heat capacity
Cp_p	Polymer heat capacity
Cp_r	Recycle water heat capacity
F_c	Catalyst flow rate
F_w	Cooling water flow rate
F_g	Recycle gas flow rate
F_{M1}	Monomer flow rate
F_{M2}	Co-monomer flow rate
F_{M3}	Hydrogen flow rate
F_{M4}	Nitrogen flow rate
O_p	Polymer outlet rate
e	Process error
Δe	Change of process error
k_d	Deactivation rate constant
k_{p1}	Monomer propagation rate constant
k_{p2}	Co-monomer propagation rate constant
$M_r Cp_r$	Thermal capacitance of reaction vessel
P_t	Total pressure
T	Bed temperature
T_f	Feed temperature
T_{wo}	Cooling water temperature after cooling
T_g	Recycle stream temperature after cooling
P_{M1}	Monomer partial pressure

P_{M2}	Co-monomer partial pressure
P_{M3}	Hydrogen partial pressure
P_{M4}	Nitrogen partial pressure
T_{wi}	Cooling water temperature before cooling
T_{ref}	Reference temperature
T_{gi}	Recycle stream temperature before cooling

References

[1] de Camargo F, da Cunha M M, dos Santos F O V, Zacca J H Z and Jardim J 2003 *Polymer* **44** 1377-1384
[2] Zambare N, Soroush M, and Grady M C 2002 *AIChE J.* **48** 1022-1033
[3] Ng C W and Hussain M A 2004 *Chem. Eng. Process* **43** 559-570
[4] Wei N C, Hussain M A and Wahab A K A 2007 *Can. J. Chem. Eng.* **85** 936-945
[5] BenAmor S, Doyle Iii F J and McFarlane R 2004 *J. Process Control* **14** 349-364
[6] Gentric C, Pla F, Latifi M A and Corriou J P 1999 *Chem. Eng. J.* **75** 131-46
[7] Vicente M, BenAmor S, Gugliotta L M, Leiza J R and Asua J M 2000 *Ind. Eng. Chem. Res.* **40** 218-227
[8] Mohd Ali J, Hoang N H, Hussain M A and Dochain D 2015 *Comput. Chem. Eng.* **76** 27-41
[9] Mohd Ali J, Hussain M A, Tade M O and Zhang J 2015 *Expert Syst. Appl.* **42** 5915-5931
[10] Spurgeon S K 2008 *Int. J. of Syst. Sci.* **39** 751-764
[11] McAuley K, MacGregor J and Hamielec A E 1990 *AIChE J.* **36** 837-850
[12] McAuley K, Talbot J and Harris T 1994 *Chem. Eng. Sci.* **49** 2035-2045
[13] Mohd Ali J and Hussain M A 2013 *ASEAN J. Chem. Eng.* **2** 50-56
[14] Mohd Ali J and Hussain M A 2013 *Proc. Int. Conf. on Chemeca 2014 (Perth Australia)* 323-328

Structural, thermal and microstructural studies of the proton conductor $BaCe_{0.7}Zr_{0.1}Y_{0.05}Zn_{0.15}O_3$ for IT-SOFCs

S Hossain[*], N Radenahmad, J H Zaini, F Begum and A K Azad

Faculty of Integrated Technologies, Universiti Brunei Darussalam,
Jalan Tungku Link, Gadong BE 1410, Brunei Darussalam.

[*]Email: 14h8854@ubd.edu.bn or, shahzad_baec@yahoo.com

Abstract. The specimen of $BaCe_{0.70}Zr_{0.10}Y_{0.05}Zn_{0.15}O_3$, a perovskite-type electrolyte, has been synthesized for application in an anode-supported protonic solid oxide fuel cell by the conventional solid state reaction in air at 1200°C for 12 hours. Structural and thermal characterization has been performed using room temperature X-ray diffraction (XRD), Scanning Electron Microscopy (SEM), Thermogravimetric Analysis (TGA) and Differential Thermal Analysis (DTA). Rietveld analysis of the XRD data has been analyzed by FullProf program and confirmed the single phase of the sample with an orthorhombic crystal structure in the Pbnm space group. To understand the temperature dependent behaviour TG/DTA scan of the precursor was recorded. The TG/DTA scan was performed under constant flow of Argon which exhibits a gradual weight loss up to 900°C. The SEM image of the pellet surface of the sample shows that the sample sintered at 1200°C was dense and suitable to use as electrolyte in solid oxide fuel cells (SOFCs).

1. Introduction

Solid-oxide fuel cells (SOFCs) operate at intermediate-temperature have attracted tremendous attention globally because of their potential long-term stability and economic competitiveness for many (including housing and automotive) applications [1]–[5]. To date, the best available electrolyte material appears to be doped ceria, which potentially meets most of the requirements for fuel-cell operation below 600 °C [6]. Some significant issues still remain, however, including electronic conductivity and undefined mechanical integrity under fuel-cell operating conditions. The finding of a new solid electrolyte for low-temperature SOFCs is a great challenge for the SOFC community. Proton conductors show potential candidates as electrolytes for low-temperature SOFCs because of their low activation energy of protons conduction. Many perovskite-type oxides show high proton conductivity in a reducing atmosphere [7]–[9]. One of the most important challenges for this type of proton conductor is a proper compromise between conductivity and chemical stability [10]–[12]. As for example, doped BaCeO3 has sufficiently high ionic conductivity, but the chemical stability in a CO_2 and H_2O containing atmosphere is inadequate for fuel-cell applications.

Because $BaCeO_3$ and $BaZrO_3$ easily form solid solutions, it is possible to replace a desired fraction of Ce in $BaCeO_3$ with Zr to form a solid solution that exhibits both adequate proton conductivity as well as sufficient thermal and chemical stability over a large range of circumstances relevant to fuel cell operation [7], [13]–[15]. Compared with doped cerates, doped zirconates possess enhanced chemical stability but lower conductivity. The stability of doped barium cerates is improved by the introduction of Zr at B site in order to accomplish high proton conductivity, as well as adequate chemical and thermal stability in a broad range of conditions referring to the fuel-cell operation. However, it is complicated to get the electrolyte with high density and high sintering temperatures are always required for it. In order to improve sinterability, Babilo and Haile [16] and Tao and Irvine [17] introduced Zn into Y and Zr doped $BaCeO_3$. Another report [18] showed the stable proton-conducting electrolyte $BaCe_{0.5}Zr_{0.3}Y_{0.16}Zn_{0.04}O_{3-\delta}$ by co-doping of Zr and Zn in the lattice could be sintered densely at lower temperatures and the chemical stability was also improved. Here, we report a new composition, $BaCe_{0.7}Zr_{0.1}Y_{0.05}Zn_{0.15}O_3$ (BZCYZ), where cerium is substituted by zirconium, yttrium and zinc. The composition was prepared as a single phase perovskite at a lower temperature comparing with the other works in the literatures and investigated the structural and thermal properties.

The proton conductivity of the electrolyte is mainly dependent on the concentration of oxygen vacancies in the perovskites oxides. The oxygen vacancies in these oxides can be significantly increased by acceptor doping, i.e., doping at the B-site by lower valance elements (LVE), typically yttrium or other rare earth metals [19].

During the synthesis and the SOFC cell operation following two reactions can occur in the presence of oxygen:

$$2Ce_{Ce}^{X} + O_o^{X} + M_2O_3 \rightarrow 2M_{Ce}' + V_o'' + 2CeO_2 \quad (1)$$

$$\frac{1}{2}O_2 + V_o'' \rightarrow O_o^{X} + 2h \quad (2)$$

The following mechanism has been proposed for the incorporation of protons in a hydrogen or water rich atmosphere and their conduction [20]–[22],

$$H_2O_{(g)} + V_o'' + O_o^{X} \rightarrow 2OH_o' \quad (3)$$

The significance of this research project is to developing intermediate-temperature (500-700°C) ceramic protons conductor as an anode support and electrolyte with a lower temperature and for adequate proton conductivity to meet the commercial target of reversible SOFCs and finding compatible electrodes with optimized functionality for dual-mode operation of reversible SOFCs.

A set of tasks and associated performance objectives have been identified to meet the research goal as follows: Investigate and use solid state and wet chemical fabrication routes for $BaCe_{0.7}Zr_{0.1}Y_{0.2-x}Zn_xO_3$ ceramic proton conductors and associated electrode materials in glycine nitrate, sol-gel, and co-precipitation processes; in later case Y can be replaced by some other target materials to look at the performances.

In this work solid state reaction method was adopted to synthesis the sample since it is commonly adopted for the preparation of ceramic compounds because of its low manufacturing cost and simplicity. Many investigators have employed solid-state reactions to synthesize oxides pertinent to IT-SOFCs [17], [23].

2. Experimental

The sample was carefully synthesized as possible to avoid any of the chemical contamination and have done the other experiments with the instruments of precise conditions.

2.1. Synthesis of BCZYZ

The perovskite sample of $BaCe_{0.7}Zr_{0.1}Y_{0.05}Zn_{0.15}O_3$ of total amount 10 g was synthesized by the conventional solid state reaction in air at 1200°C for 12 hours inside a muffle furnace (box furnace, Model: ST-1700 MX-III) using the ingredients chemicals supplied from the company Sigma Aldrich. Powders of (purity more than 99.95%) $BaCO_3$, CeO_2, ZrO_2, Y_2O_3 and ZnO were used as initial chemicals in precise stoichiometric ratios. The powders were weighed individually using a precise digital micro balance (Model: AND, GR-200) and then thoroughly mixed in an agate mortar (made up of fine stone) for one hour. The sample was then milled intimately using a ball-mill machine (Paul O. Abbe Inc., Model 202421) for about 4 hours with appropriate amount of ethanol and then was dried on a magnetic heater (Stuart, Model: SB 162) with stirring. Disc-shaped pellets of 2 g were made into a

13 mm dia stainless steel die under a Hydraulic press (Hydraulic Bench Press, Kennedy, Model: HBP 010) with a pressure of 4 ton. The pellets were then fired at 900°C for 12 hours inside the muffle furnace with a heating and cooling rate of 5°C/min. Then the calcinated sample was ground in a mortar pestle for one hour and then mixed with a small amount of poly vinyl alcohol (PVA) as a binder and then ground again for about 20 minutes and finally pressed into the pellets of 2 g of 13 mm dia under the hydraulic press of 5 ton pressure. The pellets of the sample were finally sintered at 1200°C for 14 hours inside the furnace.

2.2. Structural analysis

The X-ray diffraction experiment was performed on the sample using D8 Advanced Bruker XRD system with $CuK_{\alpha 1}$ radiation of wavelength $\lambda = 1.5406$ Å in the angular range of $2\theta = 20^{\circ}$ to 80° with a step size of 0.02° at room temperature.

The programs TREOR90 [24] and Checkcell [25] were used for indexing and refinement of the lattice parameters. The XRD data were analyzed using FullProf [26] suite program.

2.3. Thermal analysis

Thermogravimetric analysis (TGA) data were collected using a SETARAM TG-DTA/DSC thermogravimetric analyzer. The samples were heated up to 900 °C in the TG furnace at a heating rate of 10 °C/ min in dry argon. TGA in pure Ar was carried out on at 10 °C/ min from 20 to 900 °C, held isothermally for 90 min, and then cooled down to 20 °C at the same rate, with flowing the same gas.

2.4. Microstructural analysis

Scanning Electron Microscopy (SEM) experiments were carried out with a high resolution SEM machine (Model: JSM-7610F, OXFORD); with SEI DETECTOR and ACCEL Voltage 5.00 Volt and the images were taken in different magnifications for getting better fracture images.

3. Results and discussion

3.1. XRD pattern analysis

Figure 1 shows the Rietveld analysis XRD spectra of $BaCe_{0.7}Zr_{0.1}Y_{0.05}Zn_{0.15}O_3$ at room temperature after sintering at 1200° C. The Rietveld analysis of the X-ray diffraction pattern confirmed the single phase perovskite structure for the sample of $BaCe_{0.7}Zr_{0.1}Y_{0.05}Zn_{0.15}O_3$ in which the peaks are identical with the BCZYZ in the literature [18]. In this XRD pattern it is shown the observed, calculated, difference between the observed and calculated data and the Bragg positions respectively.

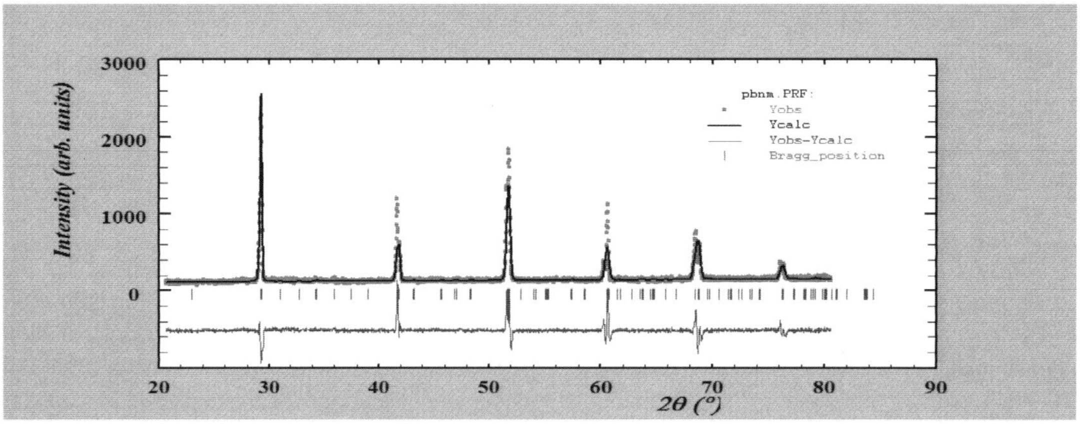

Figure 1. Rietveld analysis profile of X- ray diffraction pattern of as-prepare $BaCe_{0.7}Zr_{0.1}Y_{0.05}Zn_{0.15}O_3$ at room temperature.

The structure of $BaCe_{0.7}Zr_{0.1}Y_{0.05}Zn_{0.15}O_3$ exhibits the orthorhombic structure with the space group Pbnm and this space group has been used finally in the data refinement process for generating the calculated diffraction profiles after justified with some other crystals structures and space groups.

The refinement parameters are listed in Table 1. Figure 2 shows the schematic 3D ball and stick diagram of the material.

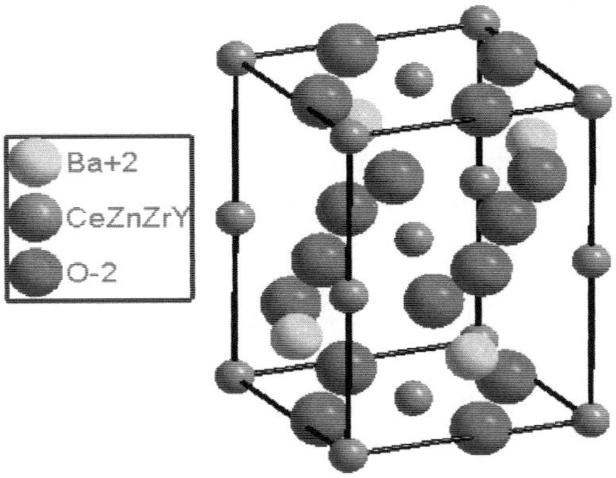

Figure 2. Schematic 3D ball and stick diagram of $BaCe_{0.7}Zr_{0.1}Y_{0.05}Zn_{0.15}O_3$.

Table 1. Crystallographic information from Rietveld analysis of the XRD data.

Quantity	Results
Unit cell parameters (Å)	a = 6.098(9), b = 6.102(7), c = 8.666(3) and $\alpha = \beta = \gamma = 90^{o}$
Space group	Pbnm
Atomic fractional coordinates	Ba (0.511,0.0120,0.250); Ce/Zr/Y/Zn (0,0,0); O1 (0.155,0.156,0.250) and O2 (0.582,0.081,0.0004)
Oxygen occupancy	98%
R-factors (%)	R_p: 11.4, R_{wp}: 16.1, R_{exp}: 7.64, Chi^2: 4.45

3.2. Scanning Electron Microscopy (SEM)

Figure 3 demonstrates the SEM of a) $BaCe_{0.7}Zr_{0.1}Y_{0.05}Zn_{0.15}O_3$ and b) the literature sample BCZY-Z within the magnifications indicated to the bottom of the pictures. The SEM micrograph of the fracture surfaces of our reported sample a) shows that the sample sintered at 1200° C were dense sample without obvious pores and crakes.

(a) (b)

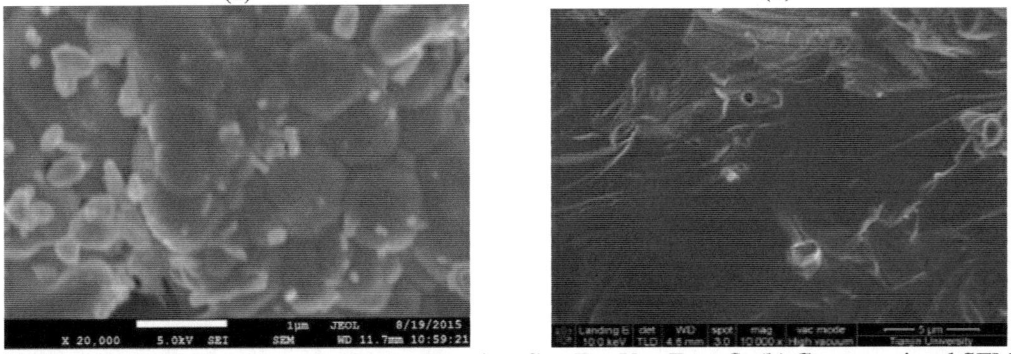

Figure 3. (a) SEM micrograph of as prepared $BaCe_{0.7}Zr_{0.1}Y_{0.05}Zn_{0.15}O_3$ (b) Cross-sectional SEM image of sample BCZY-Z of literature [27].

IOP Conf. Series: Materials Science and Engineering **121** (2016) 012014 doi:10.1088/1757-899X/121/1/012014

From the SEM image 3 a) it is clear that the grain size of the pure BCZYZ had bimodal microstructures consisting of two sizes of grains; the larger grains and other smaller grains.

The sizes of the bigger grains are 0.7-0.8 μm and smaller 0.25-0.30 μm respectively. On zinc substitution the grain size was found to be uniform. The average grain size for ceria rich compositions was larger than that for Zr rich compositions. This indicates that the nature of dopant influences the overall grain size distribution. The larger grain size of the sample offers less overall grain boundary resistance. The smaller grain size of the sample gives high overall grain boundary resistance.

Figure 3 b) represents the cross-sectional micrograph of the fracture surfaces of BCZY-Z pellets after sintering 1320° C of literature [27] which stated that the SEM of the sample confirmed a high degree of compaction, was quite dense without obvious pores and crakes. But our reported sample sintered at 1200° C shows that it is highly dense without any pores and crakes.

3.3. Thermo gravimetric analysis(TGA)

In the Figure 4 we see two curves, the blue curve shows the heat flow versus temperature graph which indicates that the heat flow was almost constant through the rising and falling during the experiment and the red one is for the weight change versus temperature which is also explained separately in Figure 5. From the Figure 5 we observe that mass is gradually decreases during the heating process and almost unchanged during the cooling of $BaCe_{0.7}Zr_{0.1}Y_{0.05}Zn_{0.15}O_3$ under the argon gas environment.

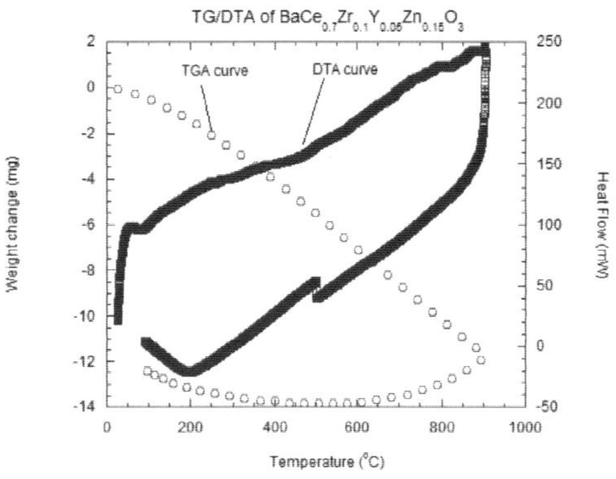

Figure 4. TG/DTA scan of $BaCe_{0.7}Zr_{0.1}Y_{0.05}Zn_{0.15}O_3$ prepared by the solid state sintering method.

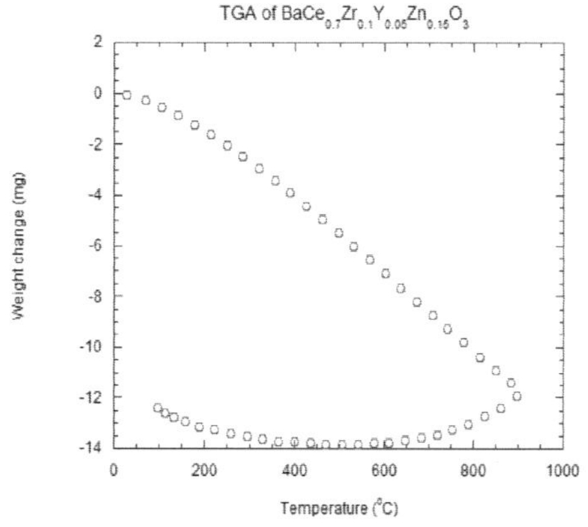

Figure 5. TGA of $BaCe_{0.7}Zr_{0.1}Y_{0.05}Zn_{0.15}O_3$ prepared by the solid state sintering method under dry argon flow.

The TGA measurement shows a two stages weight loss. In the case of argon, the initial weight loss takes place in between 100 and 200° C and the second weight loss was in between 300 and 900° C. The weight losses are related to the lattice oxygen loss from the structure. The loss of lattice oxygen at high temperature or in a reducing atmosphere provides oxygen vacancies for possible oxygen migration.

4. Conclusions

The X-ray diffraction pattern demonstrates that the $BaCe_{0.7}Zr_{0.1}Y_{0.05}Zn_{0.15}O_3$ perovskite prepared in the solid state ceramic method at a lower temperature comparative to other in literatures and are of single phase orthorhombic structure with Pbnm space group. The SEM profile picture of the specimen shows highly dense in nature which indicates that this sample can be the good electrolyte material as the proton conductor. The TG/DTA analysis of the sample indicates that the sample prepared in the solid state reactions successfully completed at this temperature which has been sintered.

5. Acknowledgments
The author is grateful to the Bangladesh Atomic Energy Commission to give study leave to perform this research. The authors are thankful to the Department of Biological & Chemical Engineering, Chalmers University of Technology for supporting the facilities for this research work done successfully.

References
[1] Fabbri E, Pergolesi D and Traversa E 2010 *Chem. Soc. Rev.* **39** no. 11 4355–4369
[2] Duan C, Tong J, Shang M, Nikodemski S, Sanders M, Ricote S, Almansoori A and O'Hayre R 2015 *Sci.* **349** no. 6254 1321–1326
[3] De Souza E C C and Muccillo R 2010 *Mater. Res.* **13** no. 3 385–394
[4] Haugsrud R and Norby T 2006 *Nat. Mater.* **5** no. 3 193–196
[5] Karnik R N 2014 *Nature* **516** no. 7530 173–175
[6] Shi Z, Sun W and Liu W 2014 *J. Power Sources* **245** 953–957
[7] Katahira K, Kohchi Y, Shimura T and Iwahara H 2000 *Solid State Ionics* **138** no. 1–2 91–98
[8] Azad A K and Irvine J T S 2008 *Solid State Ionics* **179** no. 19–20 678–682
[9] Kannan R, Singh K, Gill S, Fürstenhaupt T and Thangadurai V 2013 *Sci. Rep.* **3** 2138
[10] Lv J, Wang L, Lei D, Guo H and Kumar R V 2009 *J. Alloys Compd.* **467** no. 1–2 376–382
[11] Azad A K and Irvine J T S 2007 *Solid State Ionics* **178** no. 7–10 635–640
[12] Sawant P, Varma S, Wani B N and Bharadwaj S R 2012 *Int. J. Hydrogen Energy* **37** no. 4 3848–3856
[13] Stevenson D A, Jiang N, Buchanan R M and Henn F E G 1993 *Solid State Ionics* **62** no. 3–4 279–285
[14] Azad A K, Kruth A and Irvine J T S 2014 *Int. J. Hydrogen Energy* **39** no. 24 12804–12811
[15] Azad A K, Setsoafia D D Y, Ming L C and Petra P M I 2015 *Adv. Mater. Res.* **1098** 104–109
[16] Babilo P and Haile S M 2005 *J. Am. Ceram. Soc.* **88** no. 9 2362–2368
[17] Tao S and Irvine J T S 2006 *Adv. Mater.* **18** no. 12 1581–1584
[18] Lu X, Ding Y and Chen Y 2009 *J. Alloys Compd.* **484** no. 1–2 856–859
[19] Bonanos N 2001 *Solid State Ionics* **145** no. 1–4 265–274
[20] Glöckner R, Islam M S and Norby T 1999 *Solid State Ionics* **122** no. 1–4 145–156
[21] Ahmed I, Knee C S, Karlsson M, Eriksson S G, Henry P F, Matic A, Engberg D and Börjesson L 2008 *J. Alloys Compd.* **450** no. 1–2 103–110
[22] Azad A K and Irvine J T S 2009 *Chem. Mater.* **21** no. 2 215–222
[23] Ahmed I, Karlsson M, Eriksson S G, Ahlberg E, Knee C S, Larsson K, Azad A K, Matic A and Börjesson L 2008 *J. Am. Ceram. Soc.* **91** no. 9 3039–3044
[24] Werner P E, Eriksson L and Westdah M 1985 *J. Appl. Crystallogr.* **18** 367–370
[25] Lougier J and Bochu B ***Checkcell:*** Graphical Powder Diffraction Indexing Cell and Space Group Assignment Software.
[26] Rodriguez-Carvajal J 1993 *Phys. B Condens. Matter.* **192** no. 1–2 55-69
[27] Li Y, Guo R, Wang C, Liu Y, Shao Z, An J and Liu C 2013 *Electrochim. Acta* **95** 95–101

Synthesis and characterization of $La_{0.75}Sr_{0.25}Mn_{0.5}Cr_{0.2}Ti_{0.3}O_3$ Anodes for SOFCs

A H Karim[1, 2], A Afif[1], F Begum[1], P I Petra[1] and A K Azad[1]

[1]Faculty of Integrated Technology, Universiti Brunei Darussalam, Gadong B.E 1410, Negara Brunei Darussalam
[2]Corresponding author
Email : 14H0952@ubd.edu.bn

Abstract. $La_{0.75}Sr_{0.25}Mn_{0.5}Cr_{0.2}Ti_{0.3}O_3$ has been synthesized in solid state reaction method and tested as a potential anode material for solid oxide fuel cells. Rietveld refinement of X-ray powder diffraction data of the composition using Fullprof software shows that the materials crystallize in the rhombohedral symmetry in the R-3C space group. The cell parameters are: a = b = 5.5143 (4) Å, c = 13.452(1) Å, α = β = 90°, δ = 120°. Addition of titanium to the B-site of $La_{0.75}Sr_{0.25}Cr_{0.5}Mn_{0.5}O_{3-\delta}$ yields a total conductivity of 1.96 Scm^{-1} in air at 800 °C with activation energy of 1.02 eV.

1. Introduction

Ni-YSZ cermet is the state of the art anode material for zirconia-based solid oxide fuel cells (SOFCs). However, this material suffers some drawbacks such as carbon deposition under hydrocarbon feed and poor redox cycling [1]. Recently, mixed conducting perovskite oxides such as the lanthanum and strontium titanates or lanthanum chromites, has been given attention [2-3]. The lanthanum and strontium chromo-manganite $La_{0.75}Sr_{0.25}Cr_{0.5}Mn_{0.5}O_{3-\delta}$ were reported to be a promising anode material to replace Ni-YSZ cermet [3] and it shows good performances as a catalytic material for hydrogen and methane oxidation and redox stability in low steam to hydrocarbon ratios. Nevertheless, LSCM presents a low conductivity in air and in reducing conditions [4], and oxygen diffusion as well as surface exchange properties of LSCM [5] are lower than in other perovskites [6]. Although LSCM represents a promising anode material for SOFCs, various studies were done on LSCM based composites with yttria stabilized zirconia [7], gadolinia doped ceria [8] or metal catalyst [9,10] to improve anodic performance.

Adding dopants of transition metal elements enhances the properties of the anodes. The effects of the content and the nature of the dopant at A- or B- sites of $La_{1-x}Sr_xCr_{1-y}Mn_yO_{3-\delta}$ solid solutions on electric and electrocatalytic properties have been evaluated [11-17] to assess the potential of these oxides as SOFC anodes. For instance, 10% of cerium substituted on the A-sites of LSCM was found to enhance the electrochemical performance without additional microstructural modification and without any change of the electrical conductivity in a reducing atmosphere [18]. Knowing the chemical flexibility in perovskite structure, titanium can be partially or totally substituted for chromium in LSCM, in order to study the effect of the material basicity on its overall performance.

In this study, we have prepared and characterized a new series of materials where Cr was substituted by Ti in $La_{0.75}Sr_{0.25}Cr_{0.5}Mn_{0.5}O_{3-\delta}$. Material was prepared by solid state sintering and characterized by using X-ray diffraction, four-probe conductivity and density measurements. The characteristics were found to be suitable to use this material as anode for solid oxide fuel cells.

2. Experimental

Powders of $La_{0.75}Sr_{0.25}Cr_{0.2}Mn_{0.5}Ti_{0.3}O_3$ were synthesized by the solid state sintering method. A total of 15.311 g of $La_{0.75}Sr_{0.25}Cr_{0.2}Mn_{0.5}Ti_{0.3}O_3$ was made by adding the following powders: 8.103 g of La_2O_3 (Merck Aldrich, 99.9%), 1.718 g of SrO (Merck Aldrich, 99.9%), 2.897 g of MnO_2 (China, 99.9%), 1.588 g of TiO_2 (M erck Aldrich, 99.9%) and 1.005 g of Cr_2O_3 (Merck Aldrich, 99.9%) as according to the stochiometry. Powders were mixed with ethyl alcohol and ball milled overnight. The slurry was then put into an oven overnight at 100 °C to evaporate the solvent and the powder obtained was crushed with an agate pestle and mortar. About 2.0 g of the powder was pressed with a Kennedy hydraulic press at 3 tonnes in 20 mm radius die to make a pellet. The pellets were sintered at 1200 °C first by using a Nabertherm furnace. The pellets were again crushed and powders pelletized using the same condition and sintered at 1500 °C. In this paper, $La_{0.75}Sr_{0.25}Cr_{0.2}Mn_{0.5}Ti_{0.3}O_3$ will be identified as LSCMT.

The phase purity and crystal parameter were examined at room temperature by using a Shimadzu 7000 X-ray diffractometer using $CuK\alpha_1$ wavelength $\lambda = 0.15406$ nm at 40 kV and 30 mA and a scan speed of 1.0 deg/min. The experiment data were refined by a Rietveld method using the Fullprof [19] program. Winplotr which is available within the suit was used for powder diffraction data viewing and processing, peak profiling and indexing.

Resistance and conductivity was measured by a homemade DC four-probe setup in air over the temperature range of 100 °C to 800 °C. Ag wire and Ag paste were used to make the four probe. The electrical resistivity (ρ) was calculated by

$$\rho = RS/I \quad (1)$$

where S is the cross-sectional area and L is the length of the measured section. The conductivity (σ) was obtained by the reciprocal of the resistivity.

The pellets densities were measured by the Archimedes' method. The measured densities were compared with the theoretical densities using the relation

$$\rho_{th} = \frac{molecular\ weight\ of\ compound}{N_a.\frac{\sqrt{3}}{2}.a^2.c} \quad (2)$$

where N_a is the Avogrado constant, a, and c, are the lattice parameters for the sample.

3. Results and discussion

3.1. Structural analysis

The room temperature XRD powder patterns were indexed to know the single phase nature of the samples and to find the symmetry and cell parameters. Fig. 1 shows the X-ray powder diffraction patterns with the Rietveld refinement profile on the LSCMT sample. Unit cell parameters are related to the ideal cubic perovskite as $a \approx b \approx \sqrt{2}a_p$, $c \approx 4a_p$ (z = 4) where a_p is the cell parameter of the basic cubic perovskite $a_p \approx 3.91$Å. Figure 2 shows the schematic three-dimensional view of the compounds. Structure parameters, interatomic distances and R-factors are obtained from the Rietveld analyses. Table 1 shows the cell parameters was found to be a = b = 5.5143(4) Å, c = 13.452(1) Å for LSCMT; cell volume, v = 354.25(5) $Å^3$; oxygen occupancy = 97.5%; oxygen position in 18*e* (0.460, 0, 0.25). This is close to the cell parameter of LSCM and Ce-LSCM as reported by Tao et al.[2] and Lay et al.[18] respectively. The refinement converge in a good R-factors; Rp = 9.91% and Rwp = 14.7% and $\chi^2 = 0.86$. The high values of R-factors are due to the bad statistics of data. Since χ^2 value is very reasonable and profile matching is good, we accept the refinement.

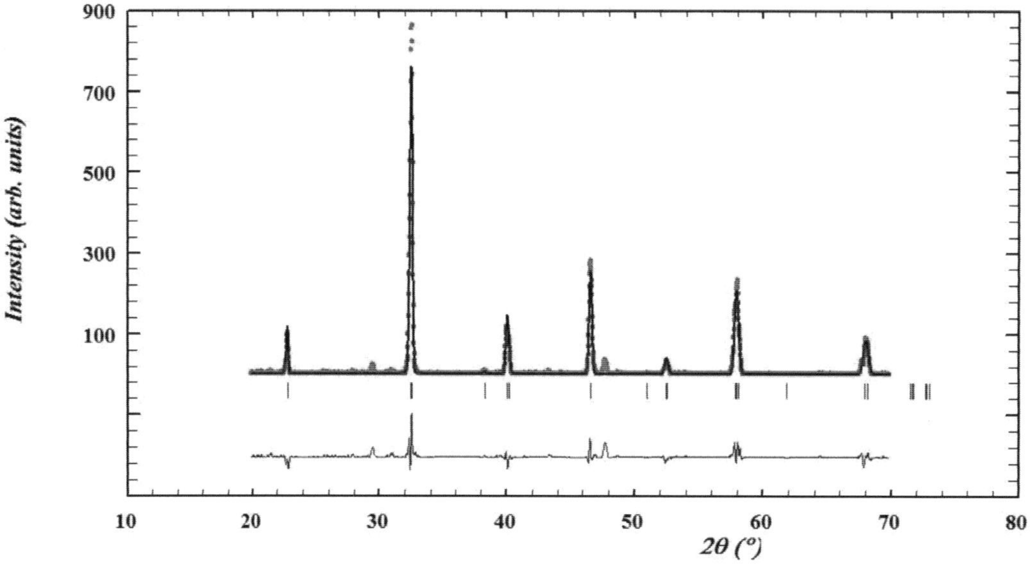

Figure 1. Observed line(red) and calculated(black) XRD intensity profiles for LSCMT. The blue line indicates the difference plot I_{obs}-I_{calc}

Table 1. Refined cell parameters for $La_{0.75}Sr_{0.25}Cr_{0.5}Mn_{0.5}O_{3-\delta}$, $La_{0.65}Ce_{0.1}Sr_{0.25}Cr_{0.5}Mn_{0.5}O_{3-\delta}$ and $La_{0.75}Sr_{0.25}Cr_{0.2}Mn_{0.5}Ti_{0.3}O_{3-\delta}$

	LSCM[2]	Ce-LSCM[18]	Ti-LSCM(this work)
a (Å)	5.5144	5.4973	5.5143
c (Å)	13.401	13.321	13.452
V (Å3)	352.92	348.65	354.25

As shown in figure 2, the B-site cations (Mn/Cr/Ti) are surrounded by regular octahedra of oxygen. These octahedra are linked together in a corner sharing three dimensional frameworks, while La/Sr ions occupy 12-coordinated A-site between these octahedra.

The Mn/Cr/Ti-O_6 octahedra are elongated along the *c*-axis. The crystal structure was distorted due to relative size differences of the A and B cations, which force the $(Mn/Cr/Ti)O_6$ octahedra to tilt in order to optimize the La/Sr-O bond distances. The tilting of the octahedra can also be realized looking the Goldschmidt tolerance factor (t) as shown in (3),

$$t = \frac{(Ra+Ro)}{\sqrt{2}(Rb+Ro)} \tag{3}$$

where Ra, Rb and Ro are the ionic radii of A-site, B-site and oxygen. When t < 1, the octahedra are tilted. Based on the ionic radii of La^{3+} (r = 1.36 Å), Sr^{2+} (r = 1.44 Å), Mn^{3+} (r = 0.645 Å),

Cr^{3+} (r = 0.615 Å), Ti^{3+} (r = 0.67 Å) and O^{2-} (r =1.40 Å) [20] in appropriate coordination number, the tolerance factor was calculated to be $t = 0.965$.

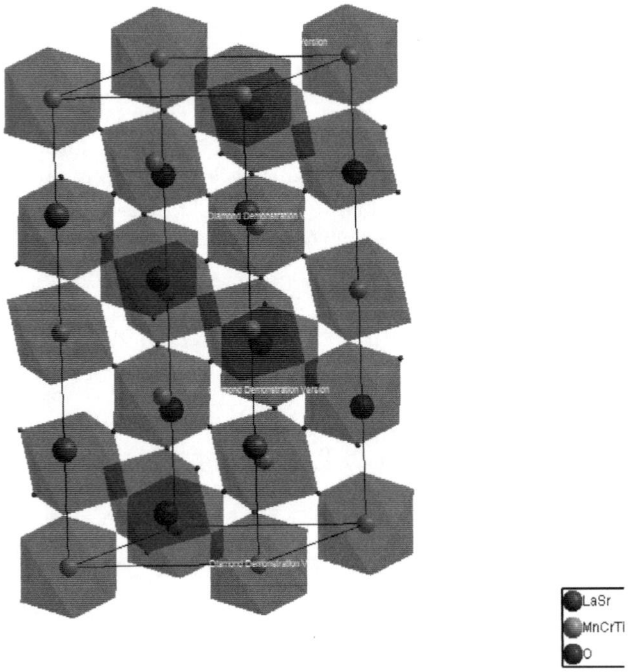

Figure 2. Schematic 3D structure of $La_{0.75}Sr_{0.25}Cr_{0.2}Mn_{0.5}Ti_{0.3}O_{3-\delta}$.

3.2 Electrical conductivity analysis

Electrical conductivity is an important parameter for SOFC anode since a sufficiently high electronic conductivity is required to provide transfer of generated charge. The minimum electrical conductivity of 1 S cm^{-1} value was proposed [1]. The Arrhenius plot of LSCMT is shown in figure 3. The maximum electrical conductivity in air 1.96 S cm^{-1}at 800 °C. The activation energy is 1.02 eV.

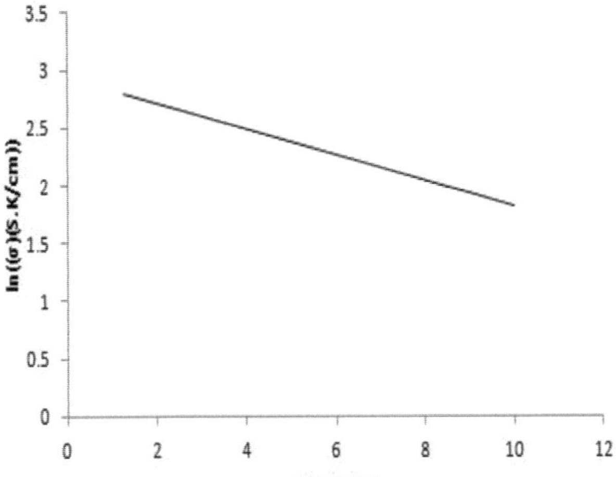

Figure 3. Electrical conductivity of LSCMT in air

3.3 Density analysis

The theoretical density is 6.44 g cm^{-3}. The actual density is found to be 6.00 g cm^{-3}. Therefore the porosity is 6.83%. This indicates that the material is suitable to be an anode for the SOFC as higher porosity is required to have a larger surface area for the anode to be in contact with hydrogen fuel during operation.

4. Conclusion

A perovskite $La_{0.75}Sr_{0.25}Mn_{0.5}Cr_{0.2}Ti_{0.3}O_3$ has been synthesized by a solid state reaction. The XRD analysis results indicate that a pure perovskite phase can be obtained in air for 8 hours and 10 hours subsequently after treatment. The crystal structure has been refined at room temperature and shows a rhombohedral symmetry in the R-3C space group. The Ti-doped material has a high electrical conductivity and low activation energy making it suitable as an anode for the SOFC.

Acknowledgments

The authors are grateful for the funding from Brunei Research Council grant S&T 17.

References

[1] Atkinson A, Barnett S, Gorte R J, Irvine J T S, McEvoy A J, Mogensen M, Singhal S V and Voh J 2004 *Nat. Mater.* **3** 17-27
[2] Tao S and Irvine J T S 2004 *J. Electrochem. Soc* **151** A252-59
[3] Canales-Vasquez J, Tao S W and Irvine J T S 2003 *Solid state ionics* **159** 159-165
[4] Plint S M, Connor P A, Tao S and Irvine J T S 2006 *Solid state ionics* **177** 2005-08
[5] Raj E S, Kilner J A and Irvine J T S 2006 *Solid state ionics* **177** 1747-52
[6] Wang L, Merkle R, Maier J, Acarturk T and Starke U 2009 *Appl. Phys. Lett.* 94
[7] Kim G, Corre G, Irvine J T S, Vohs J M and Gorte R J 2008 *Electrochem. Solid-State Lett.* **11** B16-19
[8] Chen X J, Liu Q L, Khor K A and Chan S H 2007 *J. Power Sources* **165** 34-40
[9] Zhu X, Lü Z, Wei B, Zhang Y, Liu M, Huang X and Su W 2010 *J. Electrochem. Soc.* **157** B691-96
[10] Barison S, Fabrizio M, Mortalo C, Antonucci P L, Modafferi V and Gerbasi R 2010 *Solid State Ionics* **181** 285-291
[11] Ghosh A, Azad A K and Irvine J T S 2011 *ECS Transac.* **35** 1337-43
[12] Azad A K, Hakem A and Petra P I 2015 *AIP Conference proceedings*, **1660** 070069.
[13] Azad A K and Irvine J T S 2011 *Solid State Ionics* **192** 225-228
[14] Basheer F, Ming L C, Petra P I and Azad A K 2015 *ECS Transactions* **68(1)** 1481-88
[15] Van Den Bossche M and McIntosh S 2010 *Chem. Mater.* **22** 5856-65
[16] Fonseca F C, Mucillo E N S, Mucillo R and De Florio D Z 2008 *J. Electrochem. Soc.* **155** B483-87
[17] Jeong J, Azad A K, Schlegl H, Kim B, Baek B W, Kim K, Kang H and Kim J H 2015 *Solid State Chem.* **226** 154-163
[18] Lay E, Gauthier G, Rosini S, Savaniu C and Irvine J T S 2008 *Solid State Ionics* **179** 1562-66
[19] Rodriguez-Carvajal J 1993 *Phys. B Condens. Matter*, **192** 55
[20] Shannon R D 1976 *Acta Crystallogr.* **A32** 751

Adsorption and wettability study of methyl ester sulphonate on precipitated asphaltene

H E Okafor[1], Y Sukirman[2] and R Gholami[1]

[1]Curtin University Sarawak, Department of Petroleum Engineering, CDT 250, Miri, Sarawak
[2]Universiti Teknologi Malaysia, Faculty Petroleum and Renewable Energy Engineering (FPREE), Malaysia
e-mail: henry.e@curtin.edu.my

Abstract. Asphaltene precipitation from crude oil and its subsequent aggregation forms solid, which preferentially deposit on rock surfaces causing formation damage and wettability changes leading to loss of crude oil production. To resolve this problem, asphaltene inhibitor has been injected into the formation to prevent the precipitation of asphaltene. Asphaltene inhibitors that are usually employed are generally toxic and non-biodegradable. This paper presents a new environmentally friendly asphaltene inhibitor (methyl ester sulphonate), an anionic surfactant, which has excellent sorption on formation rock surfaces. Result from adsorption study validated by Langmuir and Freundlich models indicate a favourable adsorption. At low volumes injected, methyl ester sulphonate is capable of reverting oil-wet sandstone surface to water-wet surface. Biodegradability test profile shows that for concentrations of 100-5000ppm it is biodegradable by 65-80%.

1. Introduction

Asphaltene has been identified in terms of its structure as possessing a poly-aromatic and poly-cyclic rings bearing alkyl side chains attached to heteroatoms. Asphaltene is defined in terms of solubility as insoluble in n-alkanes but readily soluble in aromatic solvents. Asphaltene colloidal state in the crude oil pose no serious concern but the precipitation from crude oil and subsequent flocculation or aggregation presents a serious concern. Asphaltene precipitation from crude oil is as a result of destabilization of thermodynamic equilibrium that exist between the asphaltene-resin micelle in the crude oil [1]. Three major factors that causes destabilization of thermodynamic equilibrium that exist in the crude oil are: changes in composition, temperature and pressure [1]. Precipitation of asphaltene from crude oil is common in undersaturated light crude oil with minor amount of asphaltene of which conversely, heavy crude oil with high amount of asphaltene content do not show asphaltene precipitation problem [2].

When asphaltene precipitate it gradually starts to settle on the rock surfaces and steady build-up of precipitated asphaltene will preferentially adsorb on the rock surfaces. The mechanism responsible for this phenomenon are due to interaction between precipitated asphaltene particles and chemical compounds in minerals present in the rock, necessitated by governing factors such as charge transfer interaction, Van der Waals, steric repulsion and hydrogen bonding [3]. The effect of adsorption of asphaltene particles on rock surface causes continuous accumulation of asphaltene particles which plugs the pore throat of reservoir rock leading to formation damage and changing the wettability of the reservoir rock which affect the production of crude oil [4]. Wettability changes is a big concern as explained by [5] precipitated asphaltene presents a positive core charge which enables it to penetrate thin water films around water-wet rock, then attaching to negatively charged water-wet mineral surfaces thereby reverting their wettability from water-wet to oil-wet.

The mechanism that prevent asphaltene particle adsorption on rock surfaces involves treating the rock surfaces with an inhibitor (surfactants) which will preferentially adsorb on the rock surface, presenting an electrostatic repulsive force that keeps the formation surface from adsorbing polar asphaltene particles as shown in Figure 6. Surfactants consists of hydrophobic and hydrophilic moieties and have been found to have the ability to alter properties of surfaces and interfaces. The mechanism by which surfactants adsorb on solid rock surfaces are: ion pairing, ion exchange, hydrophobic bonding, adsorption by polarization and adsorption by dispersion forces [6]. [7] Identified four regions by which surfactants adsorb on a rock surface and identified electrostatic attraction/repulsion and hydrophobic bonding as the mechanism for surfactant adsorption.

In the oil and gas industry surfactants has been utilized for diverse application most especially in enhanced oil recovery and production chemistry processes [8] [9]. [10] conducted adsorption studies with anionic surfactant on Berea sandstone and concluded that anionic surfactants adsorption performance are promoted by higher pH, higher temperature and low salinity. [11] Affirms similar observation by conducting static and dynamic adsorption of anionic surfactant. Several adsorption studies conducted reported a type 1 isotherm adsorption for anionic surfactants [10].

Wettability studies are usually conducted by United States Bureau of Mines (USBM) method, Amott-Harvey technique and contact angle measurement. Each of these techniques has its own limitation relating to absolute measurement and heterogeneity of the rock surfaces [12]. [13] Used zeta potential measurement to study the effect of wettability changes on carbonate rocks treated with inhibitors. The inhibitor preferential adsorb on the carbonate rock surfaces thereby repelling the oil into the bulk phase. Zeta potential measurement point of inflexion is indicative of change in wettability of the carbonate rock from oil-wet to water-wet.

In this paper, a new green anionic surfactant Methyl Ester Sulphonate (MES) is presented. Static adsorption study of methyl ester sulphonate was conducted on a crushed Berea sandstone sample. Wettability study was conducted by electrical conductivity techniques by observing the point of inflexion when varying volumes of MES was titrated into precipitated asphaltene mix with crushed Berea sandstone. MES is an environmentally friendly surfactant, biodagradability test was conducted according to APHA 5210B standard method to determine biodegradation.

2. Material and method

Methyl ester Sulphonate was purchased from Chemithon and n-heptane was purchased Sigma Aldrich (purity > 99%). Crude oil used for this experiment has API gravity of 29° API and SARA components as follows: saturates (51.36 wt%), aromatic (38.16 wt%), Resins (7.15 wt%) and Asphaltene (3.33 wt%).

The Berea sandstone to be used for the adsorption study was crushed and sieved using a 125μm to ensure an approximate uniform sand particle size. Composition of crushed Berea sandstone was determined by X-Ray Fluorescence (XRF) shown in Table 1.

Table 1. Composition of Berea sandstone sample

Composition	Amount (wt.%)
Silica (SiO_2)	73.54
Alumina (Al_2O_3)	19.57
Magnesium Oxide (MgO)	0.82
Calcium Oxide (CaO)	0.65
Ferric Oxide (Fe_2O_3)	3.47
Ferrous Oxide (FeO)	0.71
Potassium Oxide (K_2O)	1.15
Others	0.09

800mg of sandstone was measured of which 100mg each was transferred to eight 50ml beakers. Known initial concentrations of methyl ester sulphonate was prepared in the range 100-5000ppm. The absorbance value measurement was conducted using a Perkin Elmer UV-Visible spectrophotometer to establish the calibration curve. 30ml of MES was transferred to each beaker and all eight beakers placed in an orbital shaker and agitated at a constant speed of 200 rpm at 23-25°C and allowed to reach 24 hours equilibrium period. At the completion of the equilibrium period, the samples are allowed to settle for 10 minutes and samples are collected, centrifuged at 3500rpm which the supernatant was carefully collected and analysed using the UV-Vis spectrophotometer. The final equilibrium concentration was obtained by re-entry measured absorbance value into the calibration curve to obtain the equilibrium concentration. The sorption efficiency and adsorption capacity or amount of MES adsorbed by sandstone was calculated as given:

$$\% \text{ Sorption Efficiency} = \frac{C_i - C_e}{C_i} \times \frac{100}{1} \tag{1}$$

$$q_e = \frac{(C_i - C_e).V}{m} \tag{2}$$

Where q_e adsorption capacity (mg/g), C_i is the initial concentration (mg/L), C_e is the final concentration (mg/L), V is the solution volume (L) and m is the mass of sand (mg).

2.1 Wettability Measurement

Wettability study was conducted by electrical conductivity measurement. Asphaltene was separated from a crude oil with colloidal instability index (CII) of 1.21 (CII>0.9). The asphaltene sample was separated by adding 1:40 by volume of crude oil sample to n-heptane. The mixture was shaken vigorously and kept in a dark place for 24 hours. After 24 hours the supernatant was decanted and the visibly seen precipitated asphaltene blended with a mixer at 5000rpm for 10 minutes, electrical conductivity of the blended asphaltene was measured using electrical conductivity meter and the result was 0.00 μS/cm. 100g of measured crushed berea sandstone was used for this experiment and treated with 50ml brine of composition: 1.5 wt% NaCl, 1 wt% MgCl and 0.5 wt% K_2SO_4. 100g of treated crushed Berea sandstone was placed in a beaker and gradually 50ml of asphaltene mixture was poured into the beaker containing sandstone sample and mixed vigorously. MES was titrated into the asphaltene-crushed sandstone slurry. For every 2ml MES titrated into the beaker conductivity measurement was recorded by inserting the electrical conductivity meter probe into the sand-asphaltene mix.

2.2 Critical Micelle Concentration measurement (CMC)

Electrical conductivity method was employed to measure CMC of methyl ester sulphonate surfactant for different concentrations using Mettler Toledo conductivity meter. Concentrations of MES was plotted against electrical conductivity and the point where the linear plot changes abruptly is indicative of CMC for methyl ester sulphonate as shown in Figure 2.

2.3 Biodegradability test

Methyl ester sulphonate is a new anionic surfactant and has been established as a new green surfactant with similar detergency efficiency as Linear Alkylaryl Sulfonates (LAS). Biodegradability of methyl ester sulponate was tested according to APHA 5210B standard method for five days Biochemical Oxygen Demand (BOD) in the presence of Nitrogen fixing bacteria (Nitrosomonas). 100-5000ppm concentrations of methyl ester sulphonate was used for this experiment.

3. Result and discussion

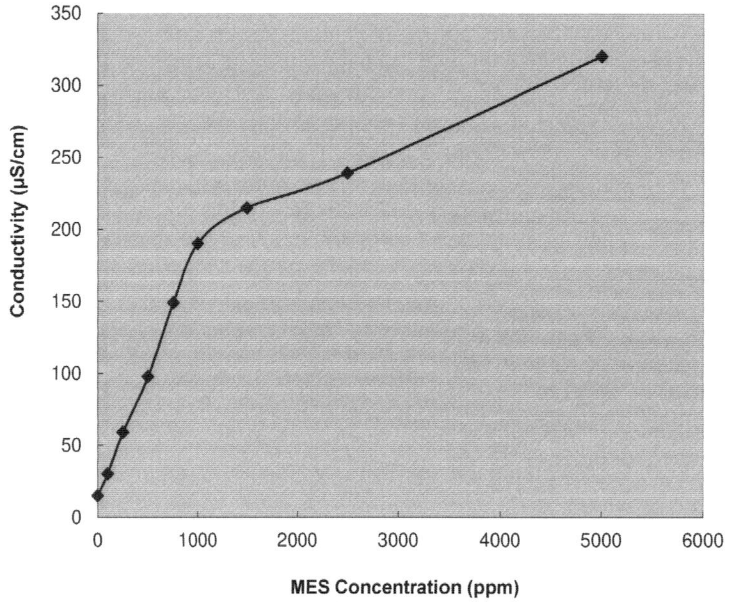

Figure 1. CMC measurement for MES by electrical conductivity

CMC measurement of MES shows an increase in the slope until and abrupt change in slope as shown in Figure 1, the point where the change was observed is known to be CMC. Thus, below the CMC MES adsorption will increase while above the CMC further increase in MES concentration does not affect adsorption. The observed CMC for MES is at 1000ppm. This point can be correlated with similar observation from adsorption isotherm shown in Figure 2 at 1000 ppm.

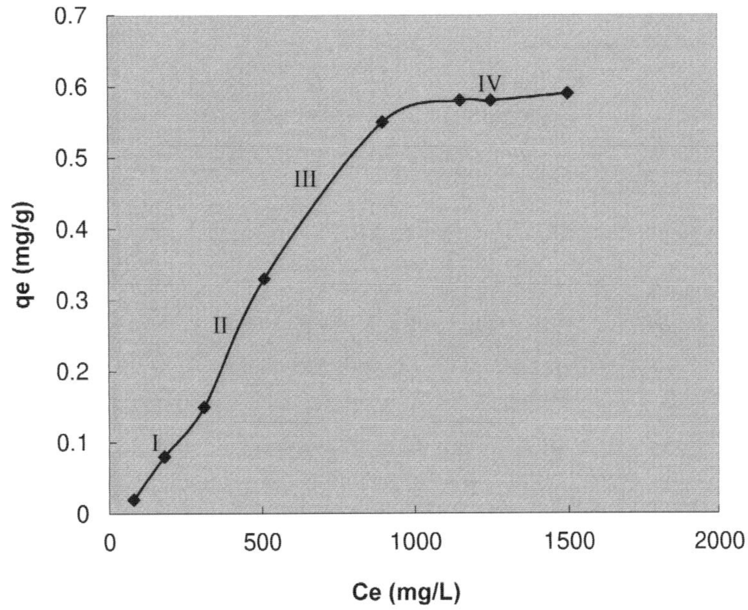

Figure 2. Adsorption isotherm of MES on Berea sandstone

Table 2. Adsorption Isotherm values for Langmuir and Fruendlich models

Langmuir			
q_m	K_L	R_L	R^2
50.5	0.0912	0.099	0.962
Fruendlich			
n	K_F	R^2	
1.158	0.017	0.982	

Langmuir adsorption isotherm describes the formation of monolayer adsorption of adsorbate on the surface of adsorbent. For Langmuir isotherm once the adsorbate occupies the surfaces, no further adsorption will occur. The Langmuir isotherm model and the dimensionless separation factor is represented as follows:

$$\frac{1}{q_e} = \frac{1}{q_m} + \frac{1}{K_L q_m} \cdot \frac{1}{C_e} \tag{3}$$

$$R_L = \frac{1}{1 + K_L C_i} \tag{4}$$

Where q_e is the adsorption capacity (mg/g), q_m is the amount of adsorbate required to form a single monolayer on unit mass of adsorbent (mg/g), K_L is the adsorption equilibrium constant (L/mg), C_e is the equilibrium constant (mg/L) and C_i is the initial concentration (mg/L). For values $0<R_L<1.0$, the adsorption is favourable and R^2 value of 0.96 shown in Table 2, also indicates that the adsorption data fits the model.

Freundlich adsorption isotherm describes the adsorption effect on heterogeneous surfaces. The Freundlich isotherm model is represented in the form:

$$q_e = K_F . C_e^n \quad (5)$$

$$\log q_e = \log K_F + n \log C_e \quad (6)$$

Where K_F is the Freundlich constant, n is the heterogeneity factor or Freundlich coefficient. For n values greater than 1.0, adsorption is termed favourable, Table 2 shows n value of 1.158 indicating a favourable adsorption. R^2 values of 0.982 is indicative of well fitted data. [14] Stated that for Freundlich isotherm model R^2 values greater than 0.89 indicates that the model can be used to validate adsorption data.

Adsorption isotherm obtained as shown in Figure 2 is similar to those reported by [7]. Four regions where depicted on the adsorption isotherm and the adsorption isotherm shows a type 1 adsorption. Region I, MES adsorption on the surfaces of Berea sandstone is predominantly by electrostatic interaction between MES head group and charged sites on the sandstone surfaces. Region II is notably marked with increase in MES adsorption resulting from electrostatic interaction occurring on the surfaces and hydrophobic interaction between adsorbed MES molecules. Region III is predominantly as a result of hydrophobic interaction and electrostatic repulsion build up. Region IV, which is above the CMC show no further adsorption as observed, as adsorption activity remains constant and a maximum adsorption of 0.59 mg/g of sandstone was recorded.

Wettability study by electrical conductivity shown in Figure 43 For the crushed Berea sandstone treated with precipitated asphaltene electrical conductivity measurement values was constant as asphaltene predominantly occupy the surface of the sandstone resulting to an oil-wet surface. Gradual addition of 2ml up to 10ml for 5000ppm MES causes conductivity value to change as MES displaces the asphaltene from the sandstone surfaces and preferentially adsorb at the surface. This occurrence reverts the wettability as MES displaces the asphaltene to occupy the surfaces thus, becoming water-wet which is important for production. The same scenario was observed for 2500ppm, 1000ppm and 500ppm and their respective point of inflexion was 18ml, 20ml and 26ml respectively. Figure 5 shows the schematics of MES displacing asphaltene by electrostatic tail-to-tail repulsion. Higher concentration of MES with less volume applied to the sample causes the reversal while higher volume of MES at low concentrations is required to effect the wettability reversal, this is as a result of fewer ion charges to interact with the charges on the surface of the sandstone.

IOP Conf. Series: Materials Science and Engineering **121** (2016) 012016 doi:10.1088/1757-899X/121/1/012016

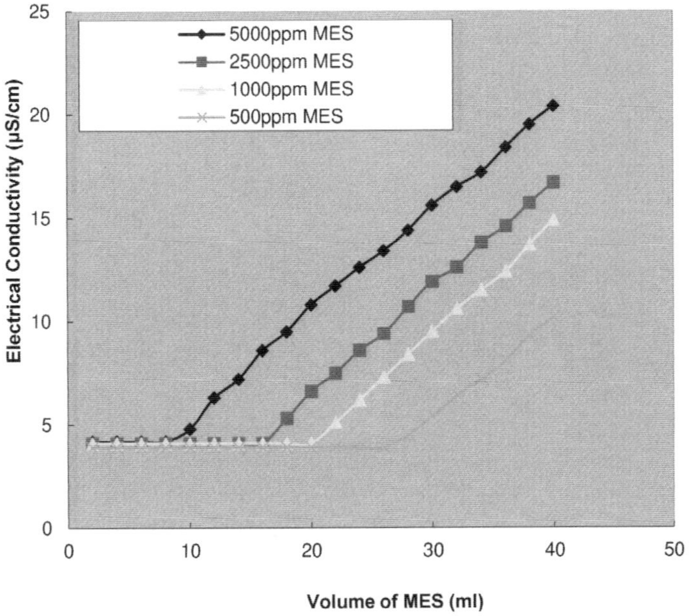

Figure 3. Wettability measurement of MES on Berea sandstone

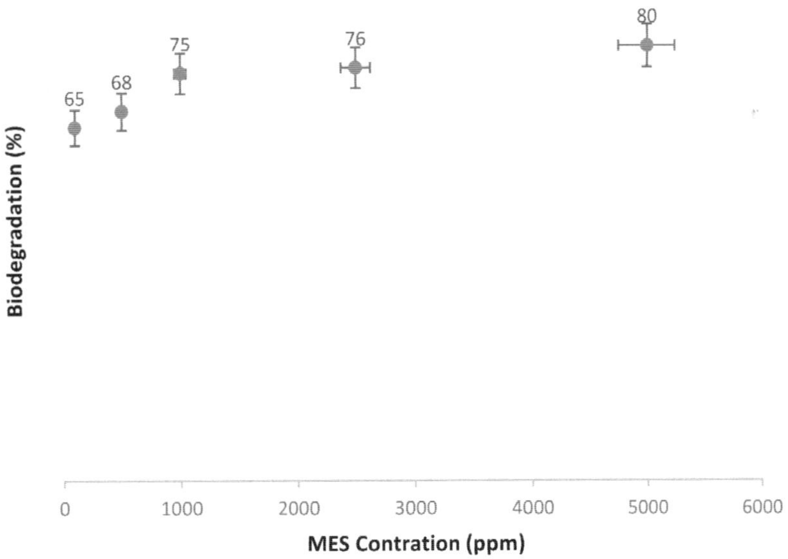

Figure 4. Biodegradation of MES Concentrations

Biodegradation test result for MES concentrations is shown in Figure 4. MES degraded in the range 65-80%, as 5000ppm MES degraded the highest by 80%. Technically, biodegradation criteria for 5 days is 60%. Results obtained meets the criteria requirement. Similar result was obtained by [15]

using different method to test the biodegradation of MES. [16] Conducted biodegradation test according to OECD 301D and reported 80% biodegradation of palm based alpha-sulphonated methyl ester an active ingredient for formulating powder detergent.

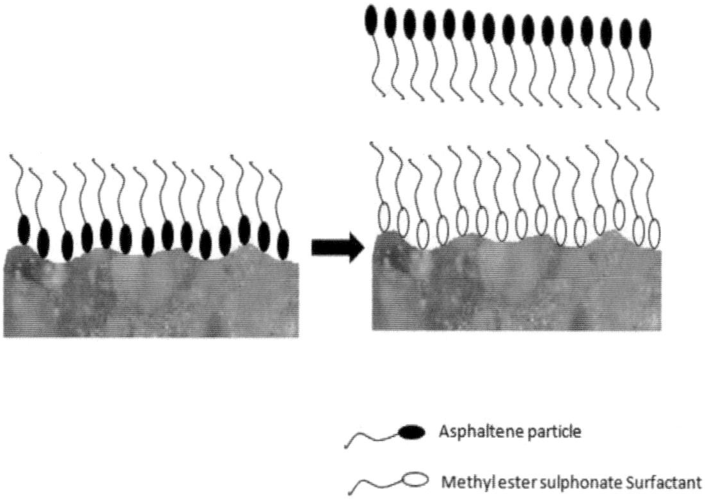

Figure 5. Schematic of MES displacing asphaltene from sandstone surface

Conclusion

Anionic surfactant methyl ester sulphonate has an excellent adsorption capacity capable of preferentially adsorbing on rock surfaces thereby reverting the state of wettability of the surface and preventing further deposition of precipitated asphaltene on the rock surfaces. As a result of MES excellent sorption capacity it can find application in chemical enhanced oil recovery processes especially in alkaline and surfactant flooding. MES is shown to be highly biodegradable, which makes it environmentally friendly surfactant.

References

[1] Leontaritis K J 1989 Asphaltene deposition: A comprehensive description of problem manifestation and modeling approaches *Society of Petroleum Engineer* 39446-MS

[2] Speight J G and Moschopedi S E 1981 On the molecular nature of petroleum asphaltenes *American Chemical Society Advances in Chemistry series* **195** pp.1-15

[3] Dubey S T and Waxman M H 1991 Asphaltene adsorption and desorption from mineral surfaces *Society of Petroleum Engineers Reservoir Engineering* pp. 389-395

[4] Leontaritis K J 1996 Asphaltene near-wellbore formation damage modeling *Society of Petroleum Engineer* 39446-MS

[5] Leontaritis K J, Amaefule J O, and Charles R E 1994 A systematic approach for the prevention and treatment of formation damage caused by asphaltene deposition *Society of Petroleum Engineers* 23810-PA

[6] Moritis G 2002 Report on enhanced oil recovery *Oil Gas Journal*

[7] Somasundra P and Zhang L 2006 Adsorption of surfactants on minerals for wettability control in improved oil recovery processes *Journal of Petroleum Science and Engineering* **52** pp. 198-212

[8] Grigg R B and Mikhalin A A 2007 Effects of flow conditions and surfactant availability on adsorption International symposium on oilfield chemistry *Society of Petroleum Engineers*

[9] Welton T D, Bryant J and Funkhouser G P 2007 Anionic surfactant gel treatment fluid *International symposium of oilfield chemistry Society of Petroleum Engineers* 105815

[10] Azam M R, Tan I M, Ismail L, Mushtaq M, Nadeem M and Sagir M 2013 Static adsorption of anionic surfactant onto crushed Berea sandstone *Journal of Petroleum Exploration Technology* **3** pp. 195-201

[11] Weifeng L V, Bazin B, Ma D, Liu Q, Han D and Wu K 2011 Static and dynamic adsorption of anionic and amphoteric surfactants with and without alkali *Journal of Petroleum Science and Engineering* **77** pp. 209-218

[12] Ghedan S, Canbaz C H, Boyd D, Mani G and Haggag M 2011 Wettability profile of a thick carbonate reservoir by the new rise in core wettability characterization method *ADIPEC Society of Petroleum Engineers* pp. 2203-2217

[13] Bassioni G and Taqvi S T 2015 Wettability studies using zeta potential measurements *Journal of Chemistry* 2015 Article ID 743179

[14] Jaynes W F and Boyd S A 1991 Hydrophobicity of siloxane surfaces in smectites as revealed by aromatic hydrocarbon adsorption from water *Clays & Clay Minerals* **39** pp. 428-438

[15] Garcia M T, Campos E, Marsal A, and Ribosa I 2008 Biodegradability and toxicity of sulphonate-based surfactants in aerobic and anaerobic aquatic environments *Water Research* **43** pp. 295-302

[16] Maurad Z A, Ghazali R, Siwayanan P, Ismail Z, and Ahmad S 2006 Alpha-sulfonated methyl ester as an active ingredients in palm-based powder detergents *Journal of surfactants and detergents* **9** pp. 161-167

Electroencephalography (EEG) Based Control in Assistive Mobile Robots: A Review

N Murali Krishnan[1], Muralindran Mariappan[2], Karthigayan Muthukaruppan[3], Mohd Hanafi Ahmad Hijazi[4], Wong Wei Kitt[5]

[1] Faculty of Business and Humanities, Curtin University, Miri, Sarawak, Malaysia
[2] Faculty of Engineering, Universiti Malaysia Sabah (UMS), Kota Kinabalu, Sabah, Malaysia
[3] School of Science and Engineering, Manipal International University, Negeri Sembilan Darul Khusus, Malaysia
[4] Faculty of Computing and Informatics, Universiti Malaysia Sabah (UMS), Kota Kinabalu, Sabah, Malaysia
[5] Faculty of Engineering and Science, Curtin University, Miri, Sarawak, Malaysia

Email: muralikrishnan.myris@gmail.com

Abstract. Recently, EEG based control in assistive robot usage has been gradually increasing in the area of biomedical field for giving quality and stress free life for disabled and elderly people. This study reviews the deployment of EGG based control in assistive robots, especially for those who in need and neurologically disabled. The main objective of this paper is to describe the methods used for (i) EEG data acquisition and signal preprocessing, (ii) feature extraction and (iii) signal classification methods. Besides that, this study presents the specific research challenges in the designing of these control systems and future research directions.

1. Introduction

Robots are with us in everyday life, and do not exist in the industry only, but also increasingly inflowing into human life. In the past few years, due to the advancement of neuroscience has led to a new capability for robots work alongside of humans. Robots can help many kinds of people, such as people who are older, people who are injured and need rehabilitation, people with severe neuromuscular disorders such as amyotrophic lateral sclerosis (ALS), brainstem stroke, cerebral palsy, and spinal cord injury (SCI) even children with autism [1]. According to the US National spinal cord injury statistical center (NSCISC) 2014 annual report, there are 276,000 people in the US who are living with SCI with approximately 12,500 new SCI cases each year [2-4]. In general, assistive robots can perform actions that benefit people with severe physical disabilities, motor diseases, paralyzed, aged or those in need, but there is a big challenge in activating the robots with conventional input methods. To address this challenge, electroencephalography (EEG) based brain controlled robots have been proposed [5, 6]. This system allows users to send comments or instructions to robots or other external devices by using brain signals, which means, the system can bypass conventional channel of communication such as muscles, nerves, speech or others which is also referred as a brain machine interface (BMI) and it creates a new non-muscular channel for relaying a person's intentions to external devices [7-9]. Apart from that, EEGs are commonly used in research because the data acquisition method is non-invasive to the research subject and it's capable of sensing changes in electrical activity in the brain on a millisecond-level [10]. EEG based robot system has produced valuable research for a variety of applications and successfully implemented in wide range of disciplines such as biology, computer science, electrical engineering, mathematics, mechanical, medicine, and, physics etc. [11-13].

BMI is an artificial intelligence (AI) system that can identify a certain set of brain signal patterns following by sequential stages which are, signal acquisition, signal preprocessing, feature extraction, signal classification, and the robotic or external device control interface [14] (refer Figure 1).

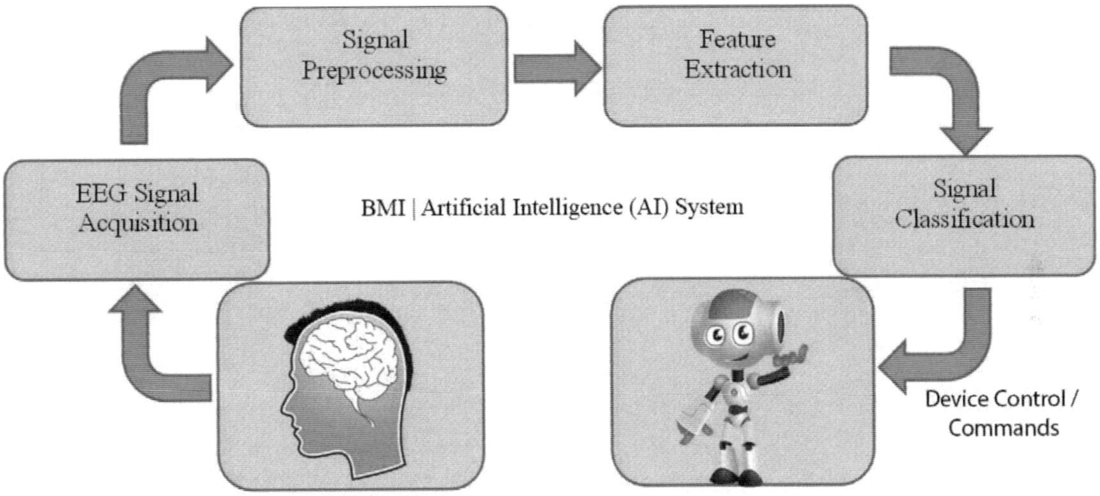

Figure 1. Stages of BMI - Artificial Intelligence

In this study, we provide a complete analysis, major techniques, and other issues of brain-controlled mobile robots along with some insights into the research and development of EEG based brain-controlled assistive mobile robots, based on our review of more than 50 related research papers.

This review has threefold; first, we present a comprehensive review of EEG signal acquisition techniques and approaches. Second, we describe and analyze the signal preprocessing, feature extraction and classification methods of brain-controlled assistive mobile robotic systems. Third, we discuss recent challenges and future directions of brain-controlled assistive mobile robot research area.

The remainder of this paper is organized as follows. In Section II, EEG signal acquisition techniques and approaches. In Section III, the signal preprocessing, feature extraction and classification methods are described and analyzed. In Section IV, we conclude this paper with the current challenges and future directions for EEG based brain-controlled assistive mobile robots.

2. Signal acquisition techniques and approaches

Direct neural interface (DNI), also called brain-machine interface (BMI) or brain-computer interface (BCI) to control assistive mobile robot is a communication system that permits direct connection between a human brain and a robot or other external device [15]. Over the past two decades, the Brain Computer Interfaces (BCI) research has developed enormously (refer Figure 2).

IOP Conf. Series: Materials Science and Engineering **121** (2016) 012017 doi:10.1088/1757-899X/121/1/012017

BCI research groups, 20 years ago, which was limited to only about three groups and there was a gradual growth after 10 years into six to eight groups.

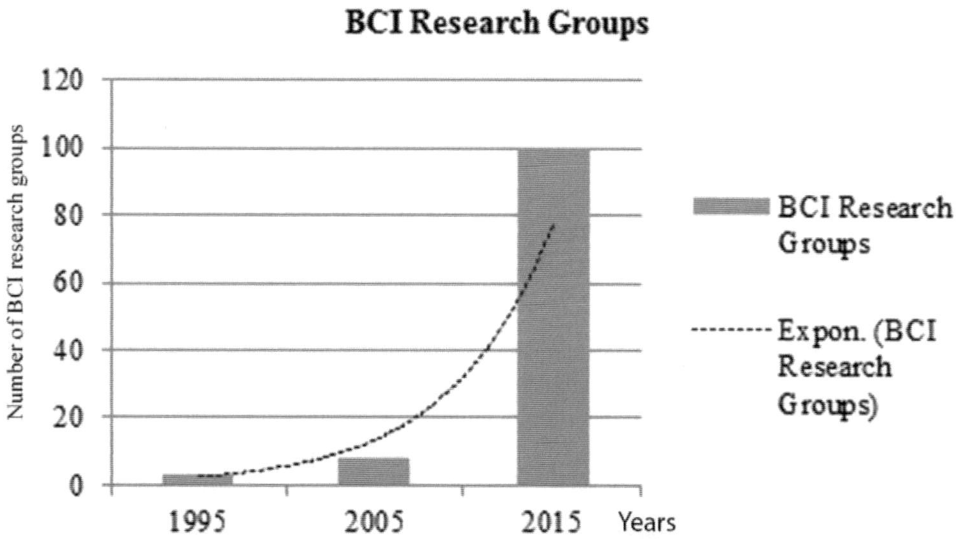

Figure 2. BCI research groups advancements

Recently, now there is a massive growth in the research groups, with over more than 100 groups throughout the world involved in an extensive range of research and development activities, and more towards in the field every month [5] [15, 16].

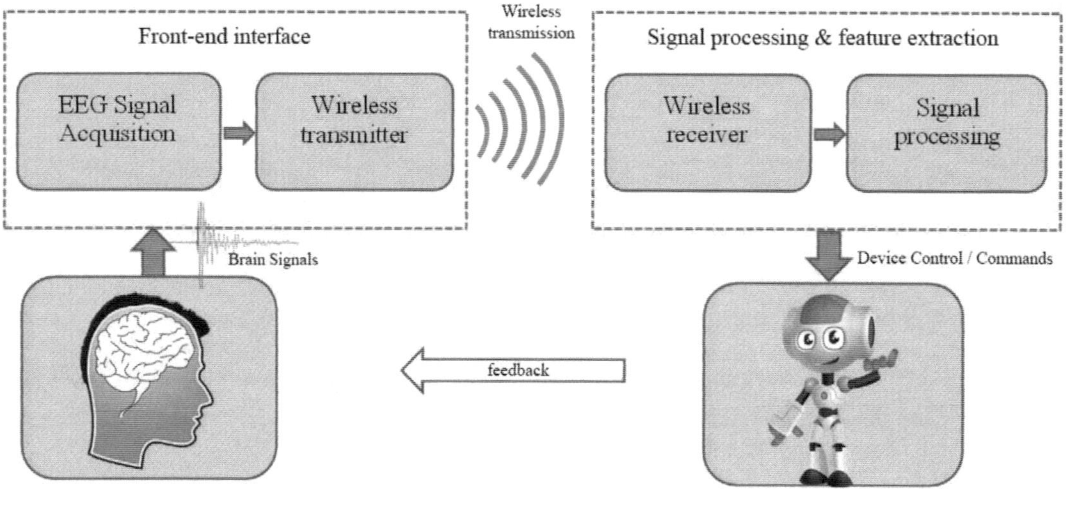

Figure 3. Components of wireless Brain Machine Interface system

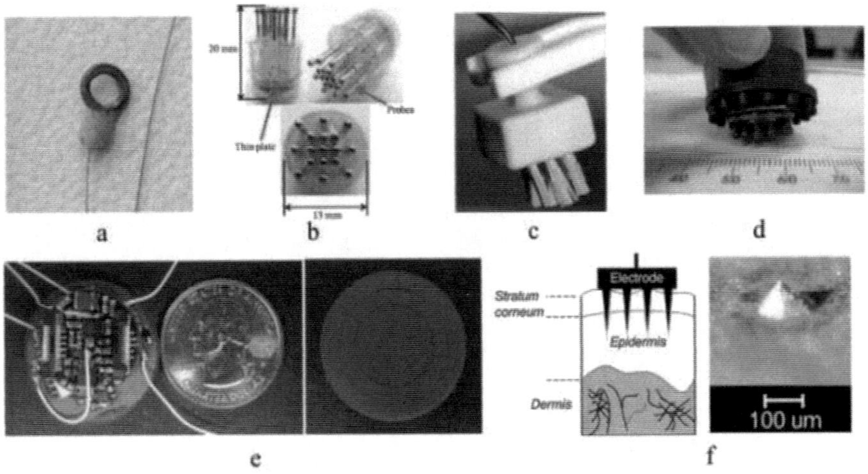

Figure 4. Various EEG signal acquisition electrode types [17]

(a) A miniature passive ring electrode (b) a spring-loaded dry electrode (c) a bristle-type dry electrode (d) the Quasar hybrid EEG biosensor (e) a non-contact-type active dry EEG sensor (f) Diagram of a micro-tip electrode and the pyramidal shape of a micro-tip

The first stage for BMI is the EEG signal acquisition, and this is the very vital session of this study. In this section contains electrodes (refer Figure 3 & 4 [17]), analog circuit and digital system for neurophysiological signal recording and transmission. The most extensively used electrodes are silver / silver chloride (Ag/AgCl) due to its stability, low contact impedance and most importantly its lower cost. The traditional BMI systems are wired, which is usually complex with a large number of cables between the electrodes and the acquisition parts. During the process of EEG data acquisition, electrodes are placed on the surface of the scalp. In the process fixing Ag/AgCl electrodes requires gel between electrodes and scalp and, this kind of electrodes are called "wet" electrodes. These processes may take longer time to fix electrodes on the scalp and which may end up in discomfort to the users. To overcome these issues, some researchers have been using "dry" electrodes [18-21]. There are few companies such as Emotiv Systems Inc., Neuro Sky Inc. have been introducing the "dry" electrodes method of non-invasive EEG data acquisition systems in the market.

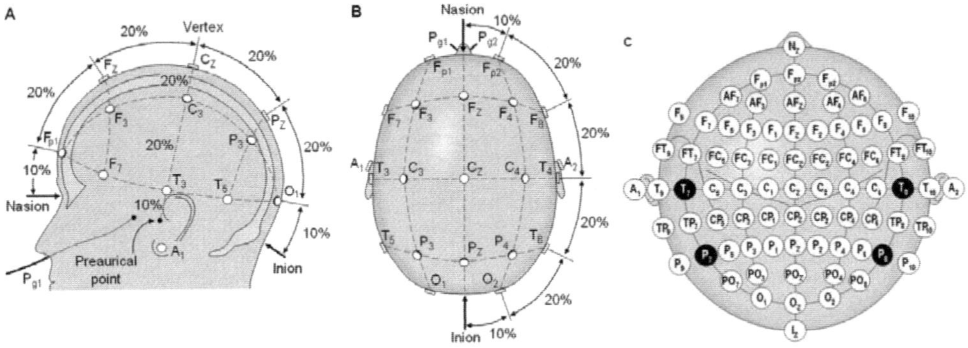

Figure 5. The international 10-20 system seen from (A) left and (B) above the head. A = Ear lobe, C = central, Pg = nasopharyngeal, P = parietal, F = frontal, Fp = frontal polar, O = occipital.

(C) Location and nomenclature of the intermediate 10% electrodes, as standardized by the American Electroencephalographic Society [22]

The subsequent significant issue of EEG acquisition is the placement of electrode positions on the scalp. In general, electrodes are placed on the scalp at 10% and 20% of a measured distance from reference sites according to the standard of 10–20 international system, as shown in Figure 5 [22].

Recently, with growing interest, wireless BCI systems have been implemented in biomedical, engineering and entertainment area as well. There are many companies recently developed their wireless EEG BMI / BCI headsets such as Emotiv and Neurosky for mind monitoring and gaming (refer Figure 6).

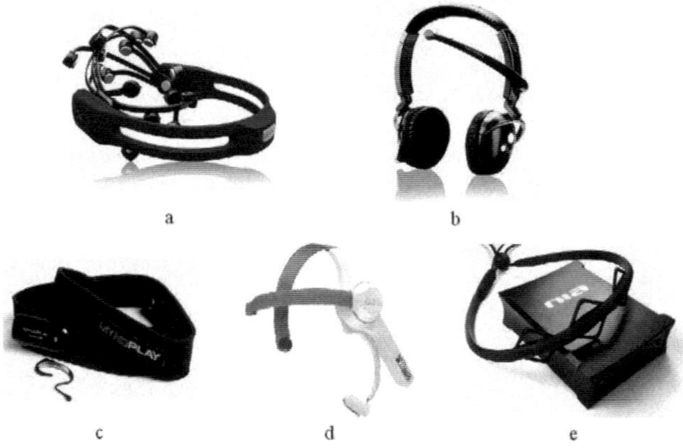

Figure 6. EEG wireless data acquisition system

(a) Emotive EPOC headset [23], (b) NeuroSky Mind Set [24], (c) MyndPlay Brainband [25], (d) PLX devices XWave headset [26], (e) OCZ Neural Impulse Actuator [27]

Furthermore, the BCI / BMI research groups have practically applied wireless BCI systems for their new research applications and development (as shown in table – 1).

Table 1. Brain Controlled Mobile Robots | Signal acquisition systems.

Product	Product from	Signal	Channels	Images
eego™mylab [28]	ANT Neuro	EEG In addition, the system's functionality can be easily extended with EOG, ECG, EMG, real-time data access and physiological sensors for respiration, temperature, skin conductance and acceleration.	32 to 256 EEG channels	

Emotiv EPOC / EPOC+ [23]	Emotiv Systems	EEG	14 (plus CMS/DRL references, P3/P4 locations)	
EEG Biosensors [24]	NeuroSky	EEG In addition, EEG Biosensors has the following features, Direct connect to dry electrode One EEG channel + Reference + Ground Extremely low-level signal detection Advanced filter with high noise immunity RAW EEG at 512Hz	One EEG channel + Reference + Ground	
BioRadio for wireless EEG [29]	BioRadio	EEG	4 channels	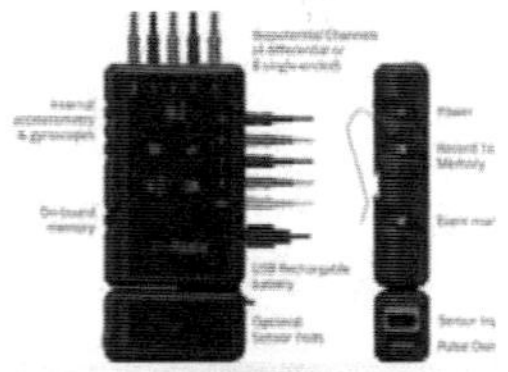
StarLab [30]	StarStim	EEG	StarStim® is a wireless hybrid EEG/tCS 8-channel neurostimulator system	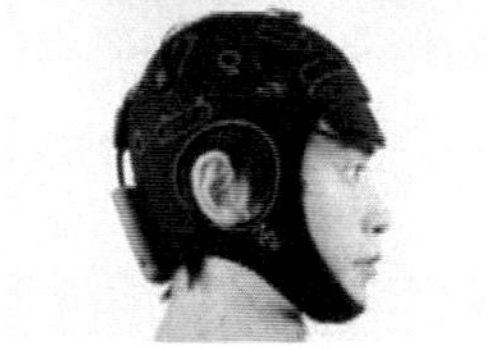
Mindo [31]	Mindo-4S Jellyfish	EGG	4 Channels	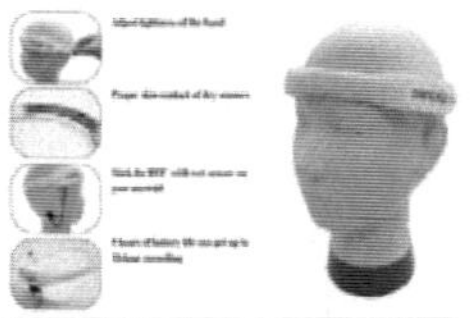

3. Signal pre-processing, feature extraction and classification

The objective of BMI or BCI is to translate the EEG signals into actions; such as by mere thinking of writing text or numbers through virtual keyboard is the perfect example of software oriented, and to navigate or monitor a mobile robot is a good example of hardware oriented. To exchange the brain signals into actions, either regression or classification algorithms can be used [32, 33]. After EEG data acquisition, the next process is signal pre-processing and feature extraction. The pre-processing stage is highly dependent on the goal of the application. There are few methods are used very commonly to improve the quality of the EEG signals, such as epoching / segmentation, filtering, artifact detection/ rejection, averaging and re-referencing. To get precise data or differentiate the required task from brain signals are apparently called as feature extraction [34] too.

Some of the feature extraction and classification techniques analyzed below the table – 2.

Table 2. Brain Controlled Mobile Robots | Signal acquisition & Classification system.

Author	Brain signal types	EEG signal acquisition	Classifier	Output commands / Application
Azar et al. [35]	Imaginary signals	EEG signals were recorded using 59 scalp electrodes	Non-linear feed-forward neural networks using the standard back-propagation algorithm	Move the wheelchair by using simple thoughts such as "go" and "stop"
P. Kumari et al. [36]	Imaginary signals	14 channels, wireless Emotiv EPOC	Wavelet transform to extract EEG signal feature and Learning Vector Quantization (LVQ) Neural Networks for classification	The EEG signal generated by visual stimuli for keying in the password.
Ting Wu et al. [37]	Imaginary signals	EEG signals were recorded using 6 channel electrodes	Probabilistic neural network (PNN) with supervised learning based on Genetic Algorithm	
Kottaimalai et al. [38]	Imaginary signals	Noninvasive type electro-cap was used for the EEG signal acquisition	Principal Component Analysis (PCA) for dimensionality decrease to eliminate the redundant variables in the EEG signals and the Neural Network as a classifier with	

			back-propagation	
Mano et al. [39]	Imaginary signals	Used invasive method and the electrodes implanted in the rat's brain	Radial basis function neural network (RBFNN) as a classifier tool for rat's brain signals	The robot was controlled by brain signal mapped into lever actions by RBFNN
Hazrati et al. [40]	Imaginary signals	Used non-invasive method	Adaptive probabilistic neural network (APNN) for EEG signal classification in an interactive online real-time BCI environment. Eye blink & eye movement signals were collected separately from the right and left earlobe channels.	Interactive online real-time BCI environment.
E. Iáñez et al. [41]	Imaginary signals	Used non-invasive method	Wavelet transform algorithm with DB2 filter has been used for signal discriminator (feature extraction), and multilayer perceptron (MLP) neural network has been used as a classifier	Robotic am controlling

Zhan et al. [42]	Imaginary signals	non-invasive electrodes	Back propagation (BP) network for the Fuzzy Neural Network (FNN) model	To move the robotic or prosthetic hand

4. Conclusion
EEG based BMI has increased attention in recent years. This study reviewed recent developments of EEG controlled mobile robots, particularly in data acquisition, feature extraction and classification area. An interesting objective of human-robot collaboration research is to build robots that are intuitive interaction partners for humans. However, more advanced solutions have to be developed for mobile robot's intelligent support system in the area of biomedical that can be used by disabled people, and thus improve their independence, mobility, and quality of life.

5. References
[1] Wolpaw J R and McFarland D J 2004 Control of a two-dimensional movement signal by a noninvasive brain-computer interface in humans *Proceedings of the National Academy of Sciences of the United States of America* **101** 17849-54
[2] NSCISC 2014 Spinal Cord Injury Model System | 2014 Annual Report Complete Public Version *NSCISC National Spinal Cord Injury Statistical Center*
[3] Boughner K J and Durfee W K 2014 Preliminary design of an energy storing orthosis for providing gait to people with spinal cord injury. In: *Engineering in Medicine and Biology Society (EMBC), 2014 36th Annual International Conference of the IEEE,* pp 2581-4
[4] Perrin X, Chavarriaga R, Colas F, Siegwart R and Millán J d R 2010 Brain-coupled interaction for semi-autonomous navigation of an assistive robot *Robot. Auton. Syst.* **58** 1246-55
[5] Millan J R, Renkens F, Mourino J and Gerstner W 2004 Noninvasive brain-actuated control of a mobile robot by human EEG *Biomedical Engineering, IEEE Transactions on* **51** 1026-33
[6] Millán J d R 2014 *Principles of Tissue Engineering (Fourth Edition),* ed R L L Vacanti (Boston: Academic Press) pp 1343-52
[7] Nicolas-Alonso L F and Gomez-Gil J 2012 Brain Computer Interfaces, a Review *Sensors (Basel, Switzerland)* **12** 1211-79
[8] Wolpaw J R, Birbaumer N, McFarland D J, Pfurtscheller G and Vaughan T M 2002 Brain–computer interfaces for communication and control *Clinical Neurophysiology* **113** 767-91
[9] Luna P 2011 Controlling machines with just the power of thought *The Lancet Neurology* **10** 780-1
[10] Plant K A, Ponnapalli P V S and Southall D M 2008 *Research and Development in Intelligent Systems XXIV,* ed M Bramer, *et al.*: Springer London) pp 363-8
[11] Millán J d R, Renkens F, Mouriño J and Gerstner W 2004 Brain-actuated interaction *Artificial Intelligence* **159** 241-59
[12] Lebedev M A and Nicolelis M A L 2006 Brain–machine interfaces: past, present and future *Trends in Neurosciences* **29** 536-46
[13] Mussa-Ivaldi F A and Miller L E 2003 Brain–machine interfaces: computational demands and clinical needs meet basic neuroscience *Trends in Neurosciences* **26** 329-34
[14] Khalid M B, Rao N I, Rizwan-i-Haque I, Munir S and Tahir F 2009 Towards a Brain Computer Interface using wavelet transform with averaged and time segmented adapted wavelets. In: *Computer, Control and Communication, 2009. IC4 2009. 2nd International Conference on,* pp 1-4
[15] Wolpaw J R 2007 Brain–computer interfaces as new brain output pathways *The Journal of*

Physiology **579** 613-9

[16] Murali Krishnan N and Mariappan M 2015 EEG-Based Brain-Machine Interface (BMI) for Controlling Mobile Robots: The Trend of Prior Studies *International Journal of Computer Science and Electronics Engineering (IJCSEE)* **3**

[17] Lin CT K L, Chang MH, Duann JR, Chen JY, Su TP, Jung TP 2010 Review of wireless and wearable electroencephalogram systems and brain-computer interfaces--a mini-review *Gerontology* **56(1):112-9. doi: 10.1159/000230807. Epub 2009 Jul 25**

[18] Sullivan T J, Deiss S R, Tzyy-Ping J and Cauwenberghs G 2008 A brain-machine interface using dry-contact, low-noise EEG sensors. In: *Circuits and Systems, 2008. ISCAS 2008. IEEE International Symposium on,* pp 1986-9

[19] Seungchan L, Younghak S and Heung-No L 2015 Design of active dry electrodes and its evaluation for EEG acquisition. In: *Information and Communication Technology Convergence (ICTC), 2015 International Conference on,* pp 560-2

[20] Yeung A, Garudadri H, Van Toen C, Mercier P, Balkan O, Makeig S and Virji-Babul N 2015 Comparison of foam-based and spring-loaded dry EEG electrodes with wet electrodes in resting and moving conditions. In: *Engineering in Medicine and Biology Society (EMBC), 2015 37th Annual International Conference of the IEEE,* pp 7131-4

[21] Kitoko V, Nguyen T N, Nguyen J S, Tran Y and Nguyen H T 2011 Performance of dry electrode with bristle in recording EEG rhythms across brain state changes. In: *Engineering in Medicine and Biology Society, EMBC, 2011 Annual International Conference of the IEEE,* pp 59-62

[22] Jaakko Malmivuo R P 1995 *Bioelectromagnetism - Principles and Applications of Bioelectric and Biomagnetic Fields*: Oxford University Press, 1995)

[23] Emotiv *http://emotiv.com/store/hardware/epoc-bci/epoc-neuroheadset/*

[24] Neurosky *http://www.neurosky.com/Products/MindSet.aspx*

[25] MyndPlay Brainband *http://myndplay.com/products.php?cat=1*

[26] PLX devices XWave headset *http://www.plxdevices.com/product_info.php?id=XWAVESONIC*

[27] OCZ Neural Impulse Actuator *http://www.ocztechnology.com/nia-game-controller.html*

[28] eego mylab *https://www.ant-neuro.com/products/eego_mylab*

[29] BioRadio *https://glneurotech.com/bioradio/physiological-signal-monitoring/wireless-eeg-research-analysis-teaching/*

[30] StarLab *http://www.starlab.es/neuroscience/*

[31] Mindo *http://mindo.com.tw/en/goods.php?act=view&no=4*

[32] Penny W D, Roberts S J, Curran E A and Stokes M J 2000 EEG-based communication: a pattern recognition approach *Rehabilitation Engineering, IEEE Transactions on* **8** 214-5

[33] McFarland D J and Wolpaw J R 2005 Sensorimotor rhythm-based brain-computer interface (BCI): feature selection by regression improves performance *Neural Systems and Rehabilitation Engineering, IEEE Transactions on* **13** 372-9

[34] Li Y and Zhang S 1996 Apply wavelet transform to analyse EEG signal. In: *Engineering in Medicine and Biology Society, 1996. Bridging Disciplines for Biomedicine. Proceedings of the 18th Annual International Conference of the IEEE,* pp 1007-8 vol.3

[35] Azar A, Balas V and Olariu T 2014 *Advanced Intelligent Computational Technologies and Decision Support Systems,* ed B Iantovics and R Kountchev: Springer International Publishing) pp 97-106

[36] Kumari P and Vaish A 2014 Brainwave based user identification system: A pilot study in robotics environment *Robotics and Autonomous Systems*

[37] Wu T, Yang B and Sun H 2010 *Life System Modeling and Intelligent Computing,* ed K Li, *et al.*: Springer Berlin Heidelberg) pp 154-62

[38] Kottaimalai R, Rajasekaran M P, Selvam V and Kannapiran B 2013 EEG signal classification using Principal Component Analysis with Neural Network in Brain Computer Interface applications. In: *Emerging Trends in Computing, Communication and Nanotechnology (ICE-CCN), 2013 International Conference on,* pp 227-31

[39] Mano M, Capi G, Tanaka N and Kawahara S 2013 An Artificial Neural Network Based Robot Controller that Uses Rat's Brain Signals *Robotics* **2** 54-65

[40] Hazrati M K and Erfanian A 2010 An online EEG-based brain–computer interface for controlling hand grasp using an adaptive probabilistic neural network *Medical Engineering & Physics* **32** 730-9

[41] Iáñez E, Furió M C, Azorín J, Huizzi J and Fernández E 2009 *Bioinspired Applications in Artificial and Natural Computation,* ed J Mira, *et al.*: Springer Berlin Heidelberg) pp 353-61

[42] Zhang X 2007 An Approach for Measurement and Recognition of Electroencephalography. In: *Electronic Measurement and Instruments, 2007. ICEMI '07. 8th International Conference on,* pp 1-698-1-702

IOP Conf. Series: Materials Science and Engineering **121** (2016) 012018 doi:10.1088/1757-899X/121/1/012018

Comparison of Two-Phase Pipe Flow in OpenFOAM with a Mechanistic Model

Adrian M Shuard[1], Hisham B Mahmud[2], Andrew J King[3]

[1] Mechanical Engineering Department, [2] Petroleum Engineering Department, Curtin University Sarawak, Miri, Malaysia

[3] Mechanical Engineering Department, Curtin University, Perth, Australia

E-mail: adrian.mark@postgrad.curtin.edu.my

Abstract. Two-phase pipe flow is a common occurrence in many industrial applications such as power generation and oil and gas transportation. Accurate prediction of liquid holdup and pressure drop is of vast importance to ensure effective design and operation of fluid transport systems. In this paper, a Computational Fluid Dynamics (CFD) study of a two-phase flow of air and water is performed using OpenFOAM. The two-phase solver, interFoam is used to identify flow patterns and generate values of liquid holdup and pressure drop, which are compared to results obtained from a two-phase mechanistic model developed by Petalas and Aziz (2002). A total of 60 simulations have been performed at three separate pipe inclinations of 0°, $+10^\circ$ and -10° respectively. A three dimensional, 0.052m diameter pipe of 4m length is used with the Shear Stress Transport (SST) $k-\omega$ turbulence model to solve the turbulent mixtures of air and water. Results show that the flow pattern behaviour and numerical values of liquid holdup and pressure drop compare reasonably well to the mechanistic model.

1. Introduction

Two-phase flow is a particular class of multiphase flow that is limited to the relative motion of two phases of immiscible fluids with different physical properties. It is common in many industrial applications such as power generation, oil and gas transportation and combustion systems. The prediction of liquid holdup and pressure drop in two-phase pipe flow is of vast importance during the design and operation stages in all applications to ensure the fluid system does not deviate from its operational envelope and cause damage to personnel, plant and equipment. Currently, steady-state hydrodynamic modelling is a technique that is widely used to help explain the behaviour and characteristics of two-phase flow [1].

Current methods for hydrodynamic modelling of two-phase flow typically comprise of either empirical or mechanistic models. Empirical models such as the well-known Beggs and Brill method [2] are an improvement of earlier homogeneous models as they provide a basis for the creation of flow regime maps and contain specific correlations for calculating liquid holdup and pressure drop. Empirical models are however based exclusively upon experimental data and are generally limited by the range of data on which they are based [3]. Alternatively, a mechanistic model is based on fundamental laws and observations and can typically offer more accurate modelling. A mechanistic transport equation is generally written for each of the phases in the system and best estimate mechanistic or correlational sub-models are used for parameters which are substituted into the equations [4]. This makes the underlying mathematics of the mechanistic model readily accessible for future amendments.

Taitel and Dukler [5] produced the first widely used two-phase mechanistic model as an analytical prediction of the transition between flow regimes in circular pipes at different inclinations. The results from the model were validated with experimental observations conducted by Mandhane et al. [6]. Many efforts have since been focused on increasing the accuracy of the analytical prediction of the flow patterns and transition boundaries such as the work of [7], [8] and [9]. The latest significant edition is the Petalas and Aziz comprehensive mechanistic model [10], which has been upgraded to apply to all pipe geometries and fluid properties. Included in the model are various amendments to parameters such as the interfacial friction factor and the effects of wall roughness and liquid entrainment have been added. The model has compared well to data from the Stanford Multiphase Flow Database [11], which consists of over 1800 experiments using actual well data and over 20,000 laboratory measurements. The high level of validation makes the Petalas and Aziz mechanistic model ideal for comparison to Computational Fluid Dynamics (CFD) models. If a good comparison is shown the models can then be used to gain a more comprehensive understanding of the flow.

The use of CFD to numerically solve flow problems has proliferated in recent years due to the advent of increased computing power and more accessible parallel computing options. Numerous examples exist of CFD analysis of two-phase flow in straight pipes however studies are usually limited to 2-dimensional analysis such as [12], or are limited in scope to a single flow regime ([13], [14]). Examples for 3-dimensional analysis of multiple flow regimes typically use different numerical models depending on the specific flow regime under investigation such as [15]. This is due to the complexity for a single solver to accurately model the behaviour of changing flow regimes. Bestion [16] summarized the challenges when attempting to use a system wide 2-phase CFD solver for multiple flow regimes when considering thermal-hydraulics of nuclear reactors. It was reported that the system wide code was typically limited to producing a macroscopic description of the flow, whereby calibration of the code is usually required if small scale effects need to be resolved accurately. In addition, the majority of CFD studies of two-phase flow have typically been attempted using the standard solvers in commercial software such as ANSYS CFX and ANSYS Fluent. In recent years OpenFOAM has gained traction as an alternative simulation tool to the common commercial software used due to the absence of licensing costs, automatic parallelisation capability and wide range of solvers available.

OpenFOAM is an open source C++ library consisting of solvers and utilities with the primary purpose for use in CFD applications. Two-phase flow is handled in OpenFOAM using interFoam, a transient solver for two incompressible, isothermal and immiscible fluids using an interface capturing technique based a modified Volume of Fluid (VOF) approach first introduced by Hirt and Nichols [17]. Validation of the interFoam solver has so far been confined to the specific interest of users and its applicability to a wide range of multiphase flow applications is yet to be established. Deshpande et al. [18] presented a summary of completed interFoam validation efforts and added further validation for pure advection, high Weber number flows and surface tension dominated flows. Any further review of the literature suggests that the use of interFoam to model two-phase pipe flow is scarce and has also been confined only to visual inspection of flow patterns based upon a specified flow map. Herreras and Labeaga [19] simulated two-phase flows of air and water in open and closed channels at different inclinations with comparison to the Taitel and Dukler flow map. Reported results however, were inaccurate due to a low mesh density and lack of a three-dimensional model. Thaker and Banerjee [20] performed a three-dimensional numerical simulation of air and water flow in a 10mm diameter, 1.5m long horizontal pipe where flow pattern results were compared by visual inspection to the experimental work of Vaze and Banerjee [21]. Results for the stratified, slug, plug and annual flow patterns compared well visually to the experimental results however no attempt was made to obtain and compare values of liquid holdup and pressure drop.

In this paper it is proposed to further compare and validate the CFD simulation results obtained using the interFoam solver by comparing values of pressure drop and liquid holdup for two-phase flow at $0°$, $+10°$ and $-10°$ inclinations to the Petalas and Aziz mechanistic model. Simulation points are selected such that comparisons can be made for the stratified wavy, slug, dispersed bubble, froth and annular flow regimes. The purpose is not to provide the most accurate analysis of each flow pattern but rather to determine whether a system wide CFD solver such as interFoam can be used with a reasonable degree of accuracy across multiple flow regimes. Successful implementation of this will enable confidence that the CFD model presented here can be utilized for further studies such as optimisation of pipeline flow or may form the basis for further studies with different fluids and pipe geometries.

2. Numerical Approach

OpenFOAM 2.1.x was used to simulate the flow using the interFoam solver. The Steady State Transport (SST) $k-\omega$ turbulence model was selected for all simulations. A three-dimensional half-pipe model was constructed with a diameter of 0.052m and a length of 4m. The geometry and mesh were both constructed in SALOME 7.4.0 and exported to OpenFOAM via the ideasUnvToFoam utility. The mesh comprised of 352628 hexahedral elements and was selected such that the boundary layer was resolved in all simulations using 8 graded layers. Values of y+ for all simulations were obtained for both phases using the yPlusRAS utility and residuals were plotted using gnuplot, after extracting the final residual data located in the log file. Gravitational acceleration was applied in the vertical (Z) direction. Analysis was conducted on different pipe inclinations by altering the appropriate vector components of the gravitational acceleration in the g file contained in the constant file directory. Fig. 1 shows the topology of the mesh.

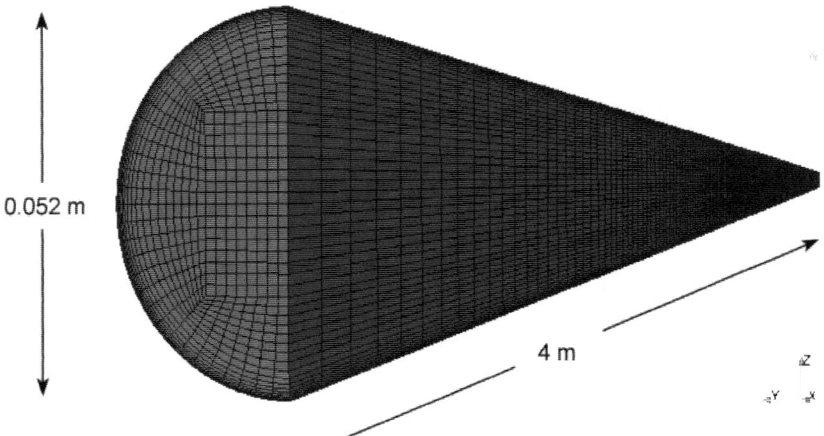

Figure 1. Mesh topology with coordinate system.

To ensure correct solution control an adjustable time step was used with an initial value of 1e-07s in conjunction with a MaxCo (mean Courant number) limit of 0.3 and a MaxAlphaCo (interface Courant number) limit set to 0.1. Standard numerical schemes for interFoam are used with the exception of *limitedLinearV 1* for the velocity component advection term to improve the solution stability. Calculations for pressure-velocity coupling are performed using the PIMPLE (merged PISO-SIMPLE) algorithm with five corrector loops and interface compression is used for the determination of phase fraction to yield a sharp interface. Other main system parameters used in the analysis are shown in Table 1.

Table 1. System parameters.

Parameter	Value
Air density	1 $[kg/m^3]$
Water density	1000 $[kg/m^3]$
Air kinematic viscosity	1.48e-05 $[m^2/s]$
Water kinematic viscosity	1e-06 $[m^2/s]$
Interfacial tension	7.24e-02 $[N/m]$
Pipe roughness (absolute)	4.57e-04 $[m]$

All results obtained from this analysis were viewed and analysed using ParaView, an open-source, multi-platform data analysis and visualisation application.

2.1. Boundary conditions

In all simulations the velocity was specified as *fixedValue* at the inlet and *zeroGradient* at the outlet. Conversely, the pressure was specified as *zeroGradient* at the inlet and *fixedValue* at the outlet. A no-slip boundary condition is applied to the pipe wall by setting the velocity as zero and a symmetry boundary condition was imposed at the longitudinal section of the half-pipe in the X-Z plane shown previously in Fig. 1. A summary of the boundary conditions used for all simulations are listed in Table 2.

Table 2. Boundary conditions.

Parameter	Inlet	Outlet	Wall
U	fixedValue	zeroGradient	fixedValue uniform 0
p_rgh	zeroGradient	fixedValue uniform 0	zeroGradient
k	fixedValue	zeroGradient	fixedValue
nut	Calculated uniform 0	Calculated uniform 0	nutkRoughWallFunction
omega	fixedValue	zeroGradient	omegaWallFunction

The pipe was initially filled with water and an equal fraction of air and water was introduced at the inlet upon commencement of the simulation (t=0s). Simulations were run until both phases had time to pass from the inlet through the entire length of the pipe. Values of liquid holdup and pressure drop were obtained from a point defined by the turbulent entrance length of each phase based on the inlet superficial velocity (whichever was greater) to a point 0.5m from the pipe outlet to reduce any inlet and outlet pressure effects.

2.2. Symmetry and mesh independency analysis

Prior to commencement of the comparative study, three simulations were run on a full 3D horizontal pipe for stratified wavy, slug and dispersed bubble flow to determine the suitability of using a symmetric boundary condition. Fig. 2a shows the longitudinal (U_x) and lateral (U_y) velocities per radial distance (sampled in the Z direction) for stratified wavy flow at the midpoint of the pipe (inlet velocities: $U_{sg} = 2$ m/s, $U_{sl} = 0.5$ m/s). The results clearly show that negligible mass flow exists across the geometric mid-plane while the behaviour of the longitudinal velocity is consistent for two-phase stratified wavy flow with a faster gas phase [20]. It has been concluded that the symmetric boundary condition applied in this study is applicable as the velocity vector across the mid-plane was also observed to be zero for all other flow patterns considered.

A mesh independency analysis was performed by comparing values of liquid holdup and pressure drop with the mesh cell count for the stratified, slug and elongated bubble flow patterns. The cell count was increased until the obtained results were less than 5% of the previous case and good agreement with the mechanistic model was achieved. Fig. 2b shows the evolution of the pressure drop depending on the total mesh cell count for a stratified wavy flow (inlet velocities: $U_{sg} = 2$ m/s, $U_{sl} = 0.5$ m/s).

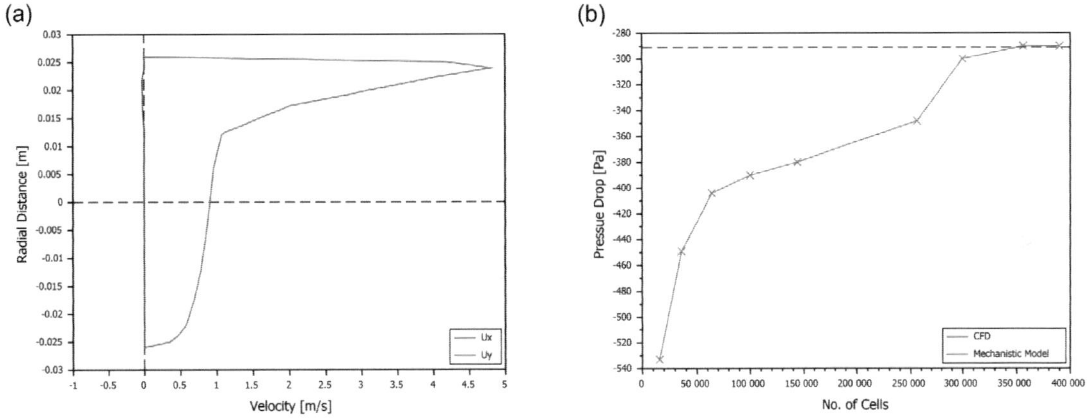

Figure 2. (a) Longitudinal and lateral velocity at the pipe midpoint for stratified wavy flow (inlet velocities: $U_{sg} = 2$ m/s, $U_{sl} = 0.5$ m/s), (b) Evolution of the pressure drop depending on the total mesh cell count for stratified wavy flow (inlet velocities: $U_{sg} = 2$ m/s, $U_{sl} = 0.5$ m/s) with a comparison to the theoretical pressure drop obtained from the mechanistic model.

2.3. Simulation points

20 simulations have been performed for each separate pipe inclination bringing the total to 60 simulations conducted. The different superficial inlet velocities for the liquid and gas phases are shown superimposed on the approximate flow regime map in Fig. 3, whereby a) is for the horizontal pipe, b) the $+10°$ inclined pipe and c) the $-10°$ declined pipe respectively. The flow regime map for each inclination has been determined by the Petalas and Aziz mechanistic model.

3. Results

3.1. Flow pattern visualisation

The general types of visualisation generated in the analysis; stratified wavy, slug, dispersed bubble, froth and annular mist flow are presented graphically in terms of volume fraction of the water phase (alpha) in Fig. 4. The blue colour represents water (alpha = 1) and the black colour represents air (alpha = 0). The interface between the two phases is represented by the white colour (alpha = 0.5).

Stratified Wavy: Results show a good visual correlation in all cases between the CFD results and the flow pattern identified by the mechanistic model. Waves form periodically at the interface and are carried downstream. An increase in the liquid inlet velocity correctly yielded

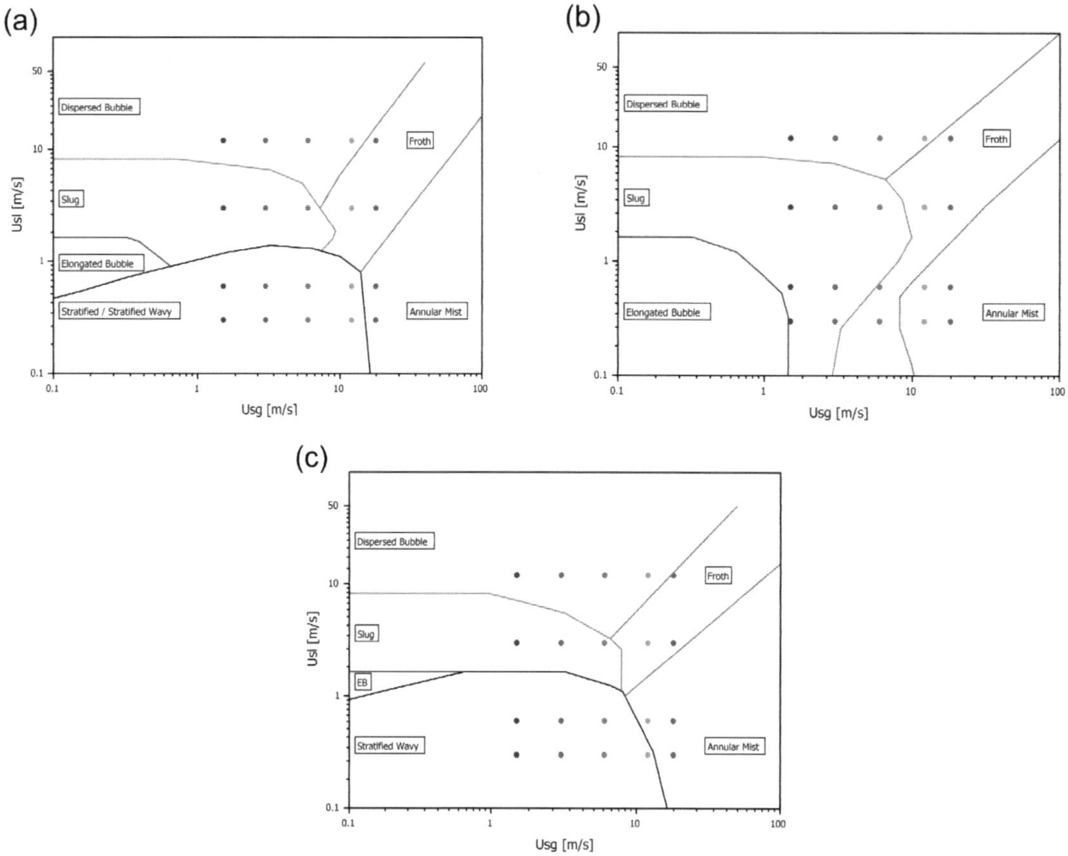

Figure 3. Petalas and Aziz flow regime maps with superimposed CFD simulation points for; (a) horizontal flow, (b) +10° (inclined) flow, (c) −10° (declined) flow.

an increase in liquid holdup while an increase in the gas velocity yielded a decrease in liquid holdup. Wave height was shown to be dependent on the velocity difference of the two phases as postulated in the Kelvin-Helmholtz wave instability theory [22]. The −10° declined pipe correctly predicted a reduction in the liquid holdup for all cases and stratified flow was correctly shown not to exist in the +10° inclined pipe.

Slug: The existence of slug flow was predicted correctly for all simulations with visualisation results closely matching numerical and experimental studies reported in the literature [23]. Slug flow was initialised by a wave forming at the interface. As the liquid velocity was greater than the stratified flow cases the liquid holdup was increased and waves that were formed were high enough to impact on the top pipe surface causing periodic slugging to occur at regular intervals. The effect of gravity was evident with slugging occurring at lower liquid velocities in the +10° inclined pipe.

Dispersed Bubble: In the dispersed bubble flow regime, the liquid height quickly reaches the top surface of the pipe whereby small groups of air bubbles were observed to be entrained in the main fluid body and carried downstream. The longitudinal mesh density was not fine

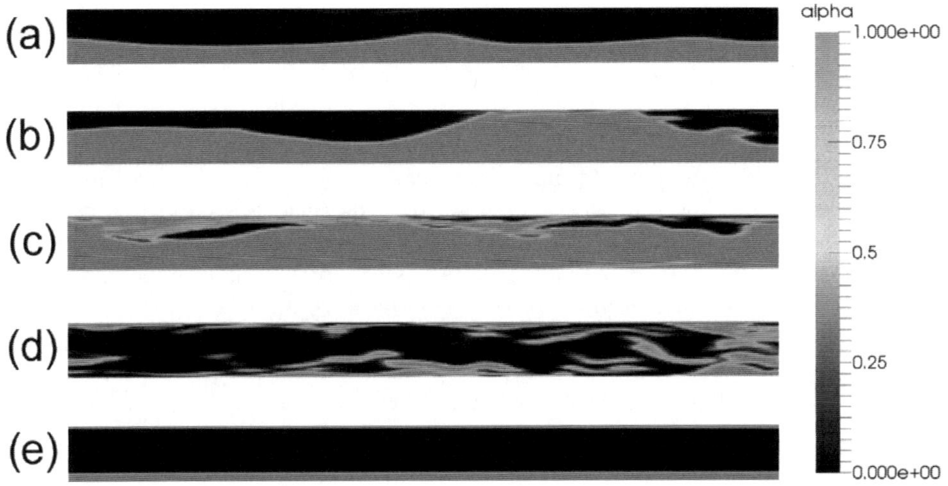

Figure 4. Flow pattern visualisation for (a) stratified wavy, (b) slug, (c) dispersed bubble, (d) froth and (e) annular mist flow.

enough to capture all the detailing on the bubbles which appear to coalesce and elongate in this study as opposed to being purely spherical as defined in the literature [24]. Pipe inclination had a negligible effect on the visualisation.

Froth: The visualisation of froth flow deviated significantly from results reported in experimental studies [25]. Froth flow comprised of wispy areas of elongated water contained within the air phase. The visual results however, were significantly different from all other flow pattern visualisations and thus can be said to match the mechanistic model predictions. Froth flow was initiated by an increased gas phase velocity which caused an initial liquid wave to quickly reach the top surface of the pipe where it was broken up and carried downstream in elongated liquid streaks.

Annular Mist: All existences of the annular flow patterns were consistent with the mechanistic model prediction. The visualisation correctly consisted of a layer of fluid at the radial extremity of the pipe with an annular core of high velocity gas; however no evidence of mist existed in the core due to the limited mesh size. The lower section of the layer was observed to be noticeably thicker due to the influence of gravity which is consistent with other experimental studies completed [26].

3.2. Liquid holdup and pressure drop comparison

Values of liquid holdup and pressure drop depending on flow regime are presented for both the CFD and mechanistic model in Figs. 5 and 6 where; a) stratified wavy, b) slug c) elongated bubble, d) froth, e) annular mist and f) represents the results from the complete set of 60 simulations with $\pm 10\%$ error margins included. The vertical lines on the plots indicate the difference between the values form the CFD analysis and mechanistic model. In this representation, the specific velocity input value for each specific phase can be deduced from the Fig. 3 flow maps whereby the simulation number from Figs. 5 and 6 represents the simulation point from Fig. 3 when read from left to right, starting from the lower left most point in the flow pattern zone. Dotted lines are used in Figs. 5 and 6 to indicate consecutive simulations that are at the same inlet liquid superficial velocity as referenced by the flow maps.

3.3. Average error magnitude
Values of the average error magnitude for liquid holdup and pressure drop depending on pipe inclination and flow pattern type are presented in Fig. 7.

4. Discussion
It is shown that the effect of changing pipe inclination is to shift the flow regime boundaries on the flow pattern map. The boundaries of flow patterns such as stratified wavy, elongated bubble, slug and froth flow are particularly influenced by changes in inclination whereas dispersed bubble and annular mist exhibit little change. The OpenFOAM simulations correctly depicted the changes in flow regimes when compared to the mechanistic model.

Stratified wavy flow was simulated for 8 cases using the horizontal pipe and 7 cases using the -10 declined pipe. A very good correlation is observed in all cases between the CFD and mechanistic model for the liquid holdup with a mean error of less than 8%, with error margins very similar for both horizontal and declined flow. A comparison of pressure drop showed good accuracy for all cases with the exception of points near the transition to annular mist flow. The effect of increasing the inlet gas velocity correctly translates to a reduction of liquid holdup and an increase in pressure drop.

Slug flow was simulated for a total of 12 cases across all three inclinations. Excellent accuracy of the liquid holdup is observed with an average error less than 6% compared to the mechanistic model with error margins similar across all inclinations. High pressure drop errors of 30% are observed in the horizontal and declined pipes whereby the increased accuracy evident in the inclined pipe is attributed to the three additional simulation points at a much lower liquid velocity which showed good correlation to the mechanistic model.

Dispersed bubble flow showed the most accurate results for the liquid holdup and pressure drop with a mean error less than 5% and 20% respectively. It is possible that this is due to the high liquid phase fraction of the flow aiding more accurate determination of the pressure drop due to a more viscous and dense combined flow. In addition, throughout all changes of pipe inclination the dispersed bubble regime is the least affected in terms of pressure drop variations. It can be seen that the flow pattern map remains roughly the same size as shown in Fig. 3. As this phase exhibits the least change it is the least affected by gravitational effects which could explain the lowest mean error.

Froth flow was simulated for 3 cases in the horizontal pipe, 5 cases for the inclined pipe and 3 cases in the declined pipe. Liquid holdup and pressure drop comparisons were generally good with less than 11% and 35% mean error respectively. Errors may be attributed to deviation in the flow pattern visualisation whereby the mesh density was not fine enough in the longitudinal direction to provide an accurate depiction of the flow.

Annular mist flow was simulated for a total of 9 cases. Results of the liquid holdup show the greatest error margin (<17%) compared to all other flow patterns considered. Mean errors in pressure drop were within 20-40% depending on the pipe inclination when compared to the mechanistic model. A good consistency is evident for liquid holdup across all inclinations with a variation of less than 1%. The highest pressure drop errors occurred in the inclined pipe and are attributed to the two additional simulation points near the flow boundary with the froth flow regime.

It can be seen that liquid holdup results are better for flows involving equal or higher flow rates for the liquid phase compared to the gas phase, with dispersed bubble flow showing the highest accuracy and annular mist flow being the least accurate. Regardless of flow regime, a good consistency in error margin is achieved between each inclination. This consistency is not immediately apparent when comparing errors in pressure drop, where large fluctuations are observed between each flow regime and inclination especially for simulation points located near flow regime boundaries. Despite this, overall it is seen that dispersed bubble flow exhibits the least mean error as in the case of the liquid holdup.

This may be explained by the weighted average values for density and viscosity used by the interFoam solver when solving the Navier-Stokes equations. The high density and viscosity ratios of air and water could be a contributing factor in introducing erroneous disturbances in the air phase at high velocities. Despite the magnitude of the pressure drop error margins recorded, the overall results give confidence that interFoam, and its implementation of the modified VOF method using the SST $k - \omega$ turbulence model can be used to give reasonable solutions to two-phase flow across multiple flow regimes, especially stratified wavy, slug and dispersed bubble.

5. Future work

It is intended to apply the same methodology presented here to a two-phase flow of oil and gas. This will yield insight into how different density and viscosity ratios could affect the flow and allow OpenFOAM to be applied to more oil and gas specific applications.

6. Conclusion

60 separate CFD simulations for two-phase flow have been completed on an air/water mixture at three separate inclinations using the interFoam solver with comparison to the Petalas and Aziz mechanistic model. All flow patterns observed in the CFD analysis were consistent with the mechanistic model prediction; however dispersed bubble and froth flow deviated from observations reported in the literature due to the limited mesh density in the longitudinal direction. A good correlation exists for the values of liquid holdup, with the highest average error occurring in the annular-mist flow regime while dispersed bubble flow exhibited the best result correlation. Similarly, the average error in the pressure drop was also observed to be highest in the annular flow regime and least in the dispersed bubble flow regime. The following conclusions are made following the results of this study;

- interFoam as a system wide solver is capable of capturing changes in the various flow patterns with a reasonable degree of accuracy for liquid holdup and pressure drop.
- Changes in pipe inclination were accurately reflected in the CFD simulation.
- Greater accuracy for both liquid holdup and pressure drop is obtained in the presence of a high velocity liquid phase as opposed to a high velocity gas phase.
- Mesh density in the longitudinal direction must be finer to provide a more accurate visualisation of dispersed bubble and froth flow.
- Usage of the SST $k - \omega$ turbulence model with a resolved boundary layer is appropriate to describe the flow with the boundary conditions used.

Nomenclature

k	turbulence kinetic energy $[m^2/s^2]$
nut	turbulence viscosity $[m^2/s]$
omega	specific rate of dissipation $[1/s]$
p_rgh	static pressure minus dynamic pressure $[kg/m - s^2]$
U	velocity $[m/s]$
U_{sl}	superficial liquid velocity $[m/s]$
U_{sg}	superficial gas velocity $[m/s]$
U_x	velocity in the X (longitudional) direction $[m/s]$
U_y	velocity in the Y (vertical) direction $[m/s]$

References

[1] Shippen M and Bailey W J 2012 Steady-state multiphase flow - past, present, and future, with a perspective on flow assurance *Energy & Fuels* **26** pp 4145-57
[2] Beggs D H and Brill J P 1973 A study of two-phase flow in inclined pipes *J. Pet. Tech.* **25** pp 607-17
[3] Ellul I R, Saether G and Shippen M E 2004 The modeling of multiphase systems under steady-state and transient conditions - a tutorial *Proc. 36th PSIG Annual Meeting (Palm Springs)* PSIG-0403
[4] Finch F, Ellul I R and Gochnour R 1991 Implementation of mechanistic flow models in a practical multiphase flow simulator *Proc. 23rd PSIG Annual Meeting (Minneapolis)*
[5] Taitel Y and Dukler A E 1976 A model for predicting flow regime transitions in horizontal and near horizontal gas-liquid flow *AIChE Journal* **22** pp 47-55
[6] Mandhane J M, Gregory G A, and Aziz K 1974 A flow pattern map for gasliquid flow in horizontal pipes *Int. J. Multiphase Flow* **1** pp 537-53
[7] Barnea D, Shoham O, and Taitel Y 1982 Flow pattern transition for downward inclined two phase flow; horizontal to vertical *Chem. Eng. Sci.* **37** pp 735-40
[8] Xiao J J, Shoham O and Brill J P 1990 A comprehensive mechanistic model for two-phase flow in pipelines *Proc. SPE Ann. Tech. Conf. (New Orleans)* SPE-20631-MS
[9] Gomez L E, Shoham O, Schmidt Z, Chokshi R N, Brown A and Northug T 1999 A unified mechanistic model for steady-state two-phase flow in wellbores and pipelines *Proc. SPE Ann. Tech. Conf. (Houston)* SPE-20631-MS
[10] Petalas N and Aziz K 2000 A mechanistic model for multiphase flow in pipes *J. of Canadian Pet. Tech.* **39** pp 43-55
[11] Petalas N and Aziz K 1995 *Stanford multiphase flow database - user's manual* Stanford Uni.
[12] Desamala A B, Dasamahapatra A K, and Mandal T K 2014 Oil-water two-phase flow characteristics in horizontal pipeline - a comprehensive cfd study *Int. J. Chem., Molecular, Nuclear, Mat. and Metallurgical Eng.* **8** pp 360-364
[13] DeSampaio P A B, Faccini J L H, and Su J 2008 Modelling of stratified gas–liquid two-phase flow in horizontal circular pipes *Int. J. of Heat and Mass Trans.* **51** pp 2752-61
[14] Rzehak R and Kriebitzsch S 2015 Multiphase cfd-simulation of bubbly pipe flow: a code comparison *Int. J. of Multiphase Flow* **68** pp 135-52
[15] Vasquez F, Stanko M, Vasquez A, DeAndrate J, and Asuaje M 2012 *Advances in Fluid Mechanics IX* ed M Rahman and C A Brebbia (Southhampton: WIT Press) pp 381-92
[16] Bestion D 2014 The difficult challenge of a two-phase cfd modelling for all flow regimes *Nuclear Eng. and Des.* **279** pp 116-25
[17] Hirt C W and Nichols B D 1981 Volume of fluid (vof) method for the dynamics of free boundaries *J. Comp. Phy.* **39** pp 201-25
[18] Deshpande S S, Anumolu L, and Trujillo M F 2012 Evaluating the performance of the two-phase flow solver interfoam *Comp. Sci. & Dis.* **5** pp 1-36
[19] Labeaga J I and Omagogeascoa N H 2013 Two-phase pipeflow simulations with openfoam *MS thesis, Norwegian Uni. of Sci. Tech. (Trondheim)*
[20] Thaker J P and Banerjee J 2013 CFD simulation of two-phase flow phenomena in horizontal pipelines using openfoam *Proc. 40th Nat. Conf. on Fluid Mech. & Fluid Power (Hamipur)* FMFP2013 paper 34
[21] Vaze M J and Banerjee J 2011 Experimental visualization of two-phase flow patterns and transition from stratified to slug flow *J. Mech. Eng. Sci.* **225** pp 382-89
[22] Yih C S 1969 *Fluid Mechanics* (New York: McGraw-Hill) p 662
[23] Vallée C, Höhne T, Prasser H M, and Sühnel T 2008 Experimental investigation and cfd simulation of horizontal stratified two-phase flow phenomena *Nuc. Eng. and Des.* **238** pp 637-46
[24] Andreussi P, Paglianti A, and Silva F S 1999 Dispersed bubble flow in horizontal pipes *Chem. Eng. Sci.* **54** pp 1101-07
[25] Gajbhiye R N and Kam S I 2011 Characterization of foam flow in horizontal pipes by using two-flow-regime concept *Chem. Eng. Sci.* **66** pp 1536-49
[26] Kumar K, Thaker J P, and Banerjee J 2013 Experimental investigations on two-phase flow phenomena in horizontal pipe *Proc. 40th Nat. Conf. on Fluid Mech. & Fluid Power (Hamirpur)* FMFP2013 paper 34

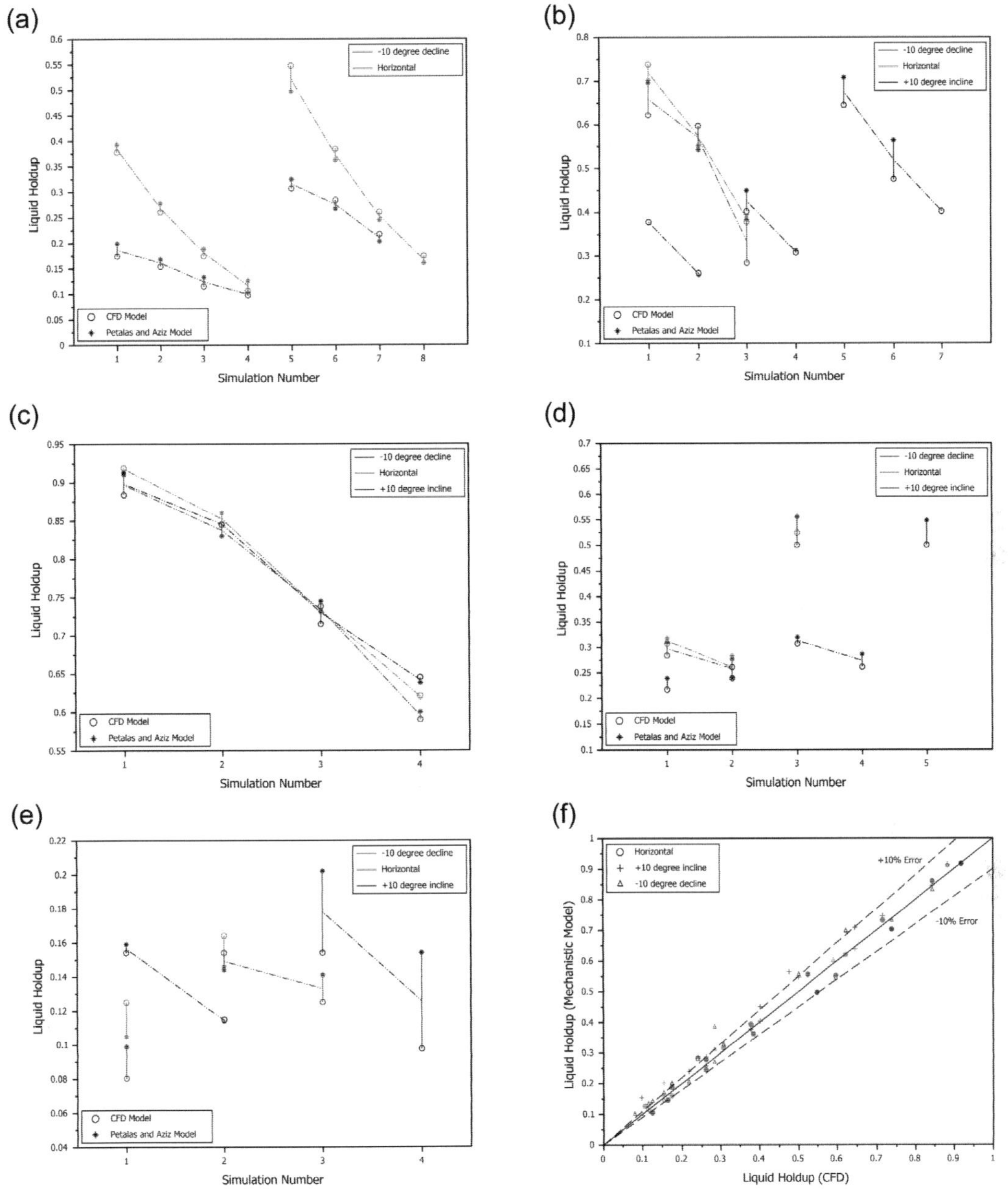

Figure 5. Liquid holdup results for each flow regime: (a) stratified wavy, (b) slug, (c) dispersed bubble, (d) froth and (e) annular mist; (f) represents a comparison of the complete data set for all inclinations with error margins shown.

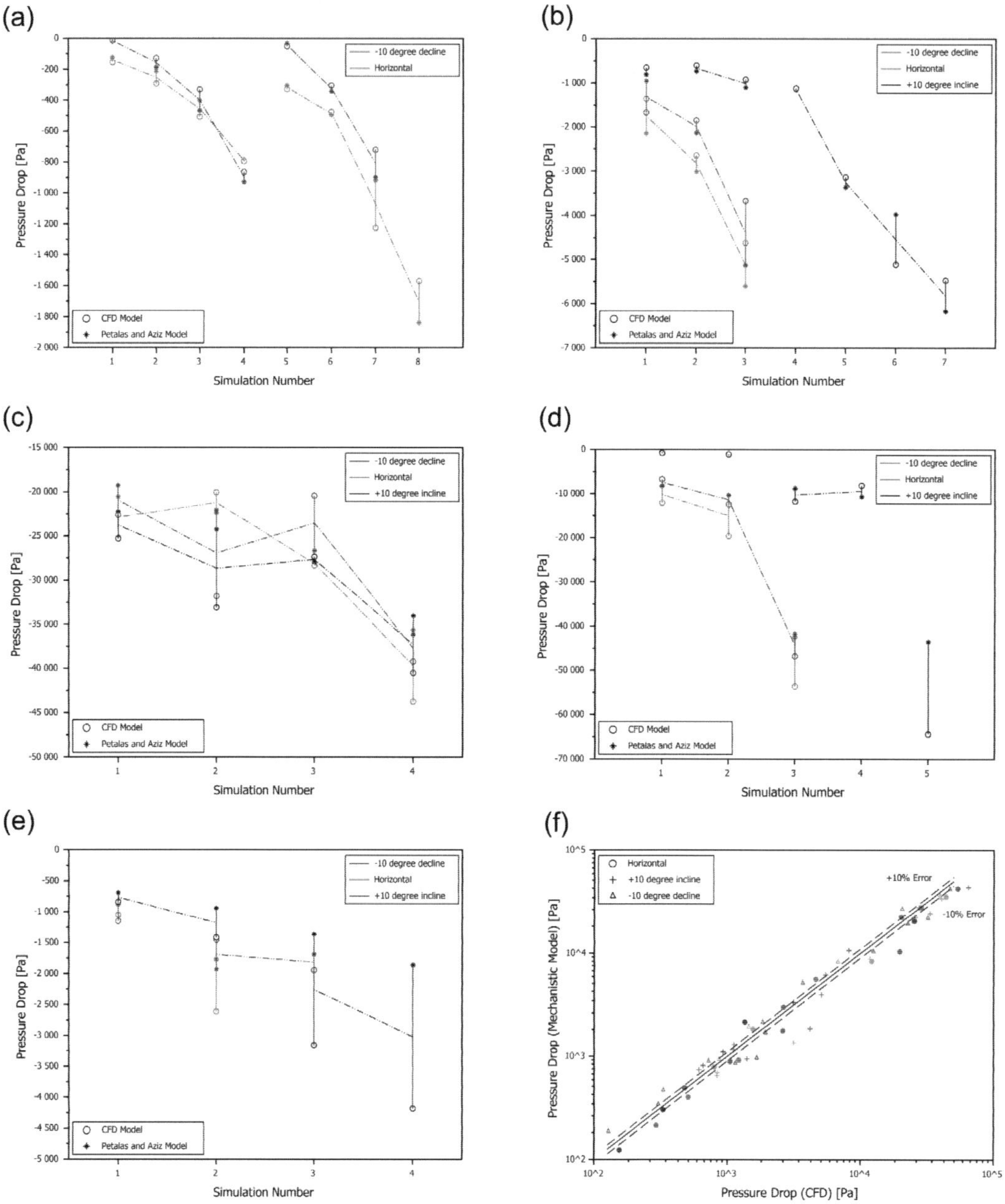

Figure 6. Pressure drop results for each flow regime: (a) stratified wavy, (b) slug, (c) dispersed bubble, (d) froth and (e) annular mist; (f) represents a comparison of the complete data set for all inclinations with error margins shown.

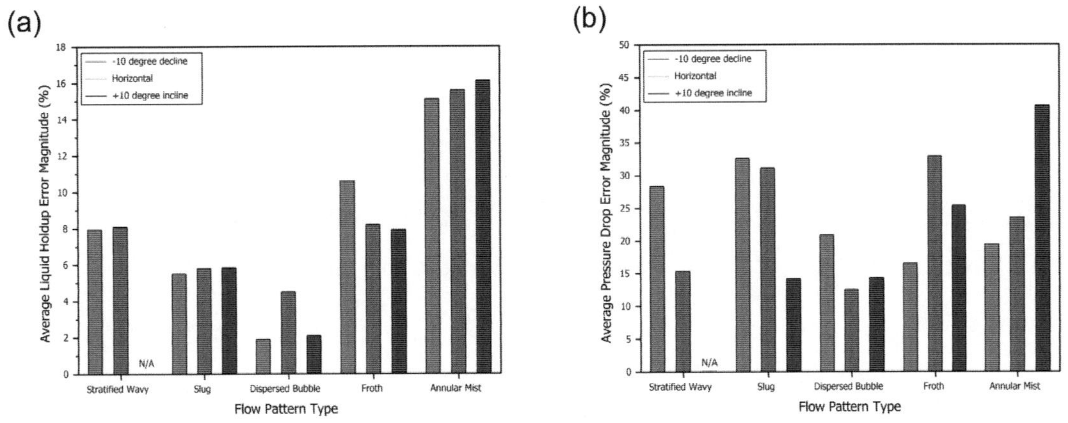

Figure 7. Average error magnitude for all flow regimes and inclinations; (a) liquid holdup and (b) pressure drop.

Water quality monitoring: A comparative case study of municipal and Curtin Sarawak's lake samples

A Anand Kumar[1], J Jaison[2, 4], K Prabakaran[1], R Nagarajan[1] and Y S Chan[3]

[1] Department of Applied Geology, Faculty of Engineering and Science, Curtin University CDT 250, 98009 Miri, Sarawak, Malaysia.
[2] Department of Chemical Engineering, Faculty of Engineering and Science, Curtin University CDT 250, 98009 Miri, Sarawak, Malaysia.
[3] Department of Petroleum Engineering, Faculty of Engineering and Science, Curtin University CDT 250, 98009 Miri, Sarawak, Malaysia.
[4] Author to whom any correspondence should be addressed.

E-mail: jaison.jeevanandam@postgrad.curtin.edu.my

Abstract. In this study, particle size distribution and zeta potential of the suspended particles in municipal water and lake surface water of Curtin Sarawak's lake were compared and the samples were analysed using dynamic light scattering method. High concentration of suspended particles affects the water quality as well as suppresses the aquatic photosynthetic systems. A new approach has been carried out in the current work to determine the particle size distribution and zeta potential of the suspended particles present in the water samples. The results for the lake samples showed that the particle size ranges from 180nm to 1345nm and the zeta potential values ranges from -8.58 mV to -26.1 mV. High zeta potential value was observed in the surface water samples of Curtin Sarawak's lake compared to the municipal water. The zeta potential values represent that the suspended particles are stable and chances of agglomeration is lower in lake water samples. Moreover, the effects of physico-chemical parameters on zeta potential of the water samples were also discussed.

1. Introduction

Surface water and ground water are important components in the environment for the fresh water resources. Photosynthetic activity of phytoplankton's and other primary producers in the aquatic environment commonly occurs in the surface water bodies [1, 2]. These planktons are the main source of food for aquatic organisms [3]. Increased in the human activities such as urban development, deforestation and discharge of industrial effluents and domestic waste into lakes from cities or villages has a profound influence in the quality of surface water bodies such as lakes and reservoirs [4, 5]. Due to the anthropogenic activities, increase in the level of suspended particles in the surface water bodies is expected [6]. The presence of suspended particles will suppress the aquatic photosynthetic system [7] and affects the aquatic food chain [8]. The term suspended particles refers to the mass or concentration of inorganic and organic particles such as trace metals [9] and carbonates [10] respectively that is present in the stream, river or lake water column by turbulence [11]. The increase of suspended particles in water bodies lead to physical and chemical alterations such as limited penetration of light which resulted in delayed photosynthesis process; temperature changes [12]; increased in eutrophication [13] and release of heavy metal contamination [14, 15] and nutrients such as phosphorus [16]. Moreover, in-situ decomposition of suspended particle with high organic content can depletes dissolved oxygen levels in water. This process will lead to a critical oxygen shortage in surface water bodies that will affect the life of aquatic organisms [17]. High level of suspended particles can also causes damage in the exposed respiratory organs of aquatic invertebrates [18] and delicate gill structures of fishes [19, 20]. Overall, the suspended particles analysis plays a major role in establishing water quality guidelines to improve and protect quality of water and aquatic ecosystems. In the present study, particle size distribution and zeta potential of the suspended particles in municipal water and surface water of Curtin Lake, Miri, East Malaysia was compared and analyzed using dynamic light scattering technique. In addition, the effect of physico-chemical parameters on zeta potential of the water samples was also analyzed.

2. Study area

The lake surface water was collected from Curtin Lake which is located in Curtin University, Sarawak campus, Miri, East Malaysia. Miri city is located in northern Sarawak, in the Borneo Island. The city is dotted by numerous industries such as palm oil, timber processing plants, ship building industries and offshore oil platforms. The surface water is available throughout the year and is utilized for domestic purposes. The Sarawak state receives more annual rainfall of about 3000-4000 mm through the northeast and southwest monsoons. Most of the rainfall is received during October to March through northeast monsoon. The south west monsoon is the dry period of Sarawak. The average temperature is 26 °C throughout the year and the state experiences higher humidity level (80%) due to high evaporation rate [21]. The burning of bushes and its aftermath contributes to the presence of major suspended particles in surface water bodies in and around Miri city. A total of 10 water samples from the surface of Curtin Lake (S1 to S10) were collected for analysis as mentioned in Figure 1. The tap water and the drinking water samples were also collected inside university campus. The tap water was supplied by local municipal authority and the drinking water was purified by Curtin University water treatment facility.

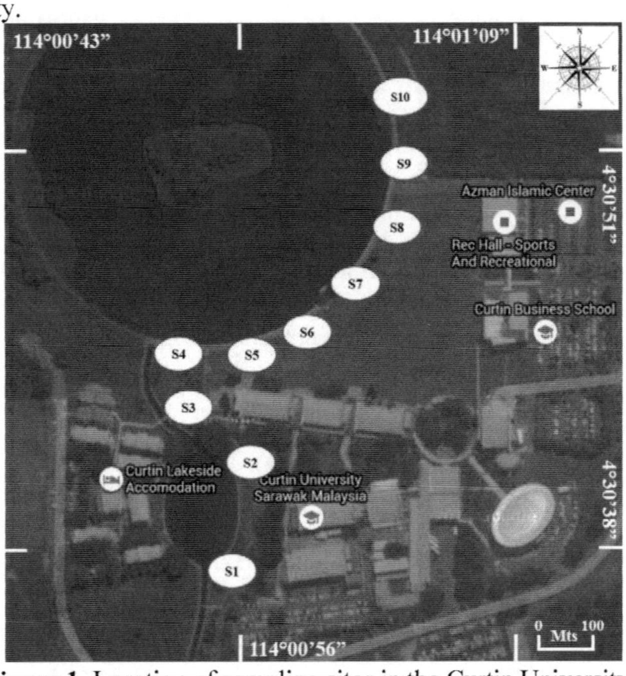

Figure 1. Location of sampling sites in the Curtin University, Sarawak Campus.

3. Materials and methods

Prior to the collection of samples the storage bottles was thoroughly acid washed with diluted Nitric acid (20%) and rinsed with distilled water. General physico-chemical parameters such as conductivity, pH, resistivity, salinity, TDS, ORP and temperature measurements of the water samples was measured in situ with portable water quality analyzer. The collected water samples were filtered using a Whatman No: 1 filter paper for further analysis. The filtered samples were taken in disposable folded capillary cells and measured by using Non-Invasive Backscatter optics (NIBS). The particle size distribution and zeta potential was analyzed using Zetasizer Nano ZS (Malvern Instruments Ltd). This instrument can able to measure the size ranging from 0.3nm o 10 microns (diameter). The accuracy and precision of the instrument was checked by using the standard reference material NIST SRM 1980 [22].

4. Materials and methods

Measured values of physico – chemical parameters, zeta potential values and particle sizes are summarized in Table 1. The range and mean values of physico-chemical parameters was represented in Table 2. The pH was found to be slightly acidic for the surface water. Electrical conductivity was observed more in the lake water samples compared to the municipal waters. Higher concentration of total dissolved solids was observed in the sampling site S1 and the minimum values was noticed in the municipal waters. Salinity of the water samples was in the range of 0.3 to 0.4 ppt. Zeta potential is a convenient way to study the suspended particle and its interactions in surface water bodies [23]. The interaction of suspended particles in aqueous media is dependent on its zeta (electro-kinetic) potential which is controlled by the surface charge of the suspended particles and the aqueous medium [24-26]. According to electric potential of colloidal particles, zeta potential from 0 to ±5mV leads to agglomeration, ±10 to ±30mV indicates incipient stability, ±30 to ±40mV leads to moderate stability and excellent stability can be obtained from ±40 to ±60mV [27, 28]. The sign and the level of zeta potential will act as a deciding factor to determine the range of physical and chemical processes on the particle surface [29, 30]. Zeta potential of all the water samples was found to be negative which denotes the surface charge of the suspended particles [31, 32] and it is represented in the figure 2. For distilled water, the pH is 6.82, the average size is 1345 nm and zeta –potential is -8.58. Compared with the physico-chemical parameters and Z-potential in all the samples (Table 1), it is clear that the presence of suspended particles are very low in the distilled water and it shows the lower Z-potential and larger size which will lead to agglomeration and settlement of the suspended particle [33, 34]. Similarly, compared to tap water, drinking water consists of larger size suspended particles (831nm) with lower z-potential (-12.1mV) which show there is no micro-level contaminants present in both samples. The pH of the lake samples are in the range of 5.8 and 6.8, the particle size varied from 180.3 nm to 386.6 nm and the zeta potential value varies between -13.7mV to -26.1mV.

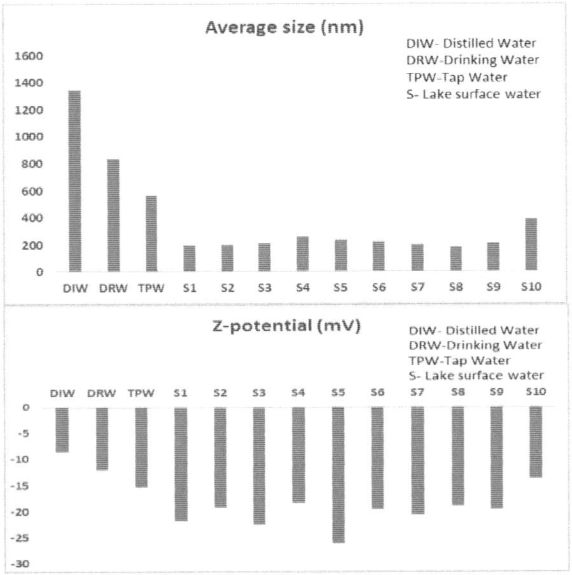

Figure 2. Chart showing zeta potential and average size of all the water samples

Table 1. **Physico-chemical parameters of water samples collected from Curtin university, Sarawak Campus**

Sample ID	pH	Conductivity (µs/cm)	Salinity (ppt)	Resistivity (MΩ-cm)	TDS (mg/l)	ORP (Rmv)	Temperature ºC	PDI	Average size (nm)	Z-potential (mV)
DIW	6.82	1.47	0.0	0.681	1	46.8	26	0.998	1345	-8.58
DRW	6.47	138.8	0.0	0.0072	68	42.7	28	0.667	831	-12.1
TPW	6.6	138.5	0.0	0.0072	68	34.3	27.7	0.537	567	-15.3
S1	6.61	850	0.4	0.0012	417	55.2	31.3	0.351	198	-21.8
S2	6.34	679	0.3	0.0015	333	49.7	30.6	0.345	198.4	-19.2
S3	6.25	632	0.3	0.0016	310	55.2	32.2	0.398	208.3	-22.5
S4	6.14	639	0.3	0.0016	313	61.7	31.8	0.423	258.2	-18.4
S5	6.13	660	0.3	0.0015	323	62.8	32	0.54	236.5	-26.1
S6	6.1	690	0.3	0.0014	338	64.1	31.5	0.339	218	-19.5
S7	6.21	631	0.3	0.0016	309	58	32.4	0.416	200.6	-20.7
S8	6.16	635	0.3	0.0016	311	60.9	32.5	0.391	180.3	-19
S9	6.15	629	0.3	0.0016	308	61.3	33	0.462	211.3	-19.6
S10	5.93	646	0.3	0.0015	317	74.3	31	0.852	386.6	-13.7

Table 2. **Descriptive analysis of water samples**

Parameters	Range	Minimum	Maximum	Mean	Std. Deviation
pH	0.89	5.93	6.82	6.30	0.25
Conductivity	848.53	1.47	850.00	536.13	261.25
Salinity	0.40	0.00	0.40	0.24	0.14
Resistivity	0.68	0.00	0.68	0.05	0.19
TDS	416.00	1.00	417.00	262.77	127.98
ORP	40.00	34.30	74.30	55.92	10.44
Temperature	7.00	26.00	33.00	30.77	2.16
Average Size	1164.70	180.30	1345.00	387.63	344.80
Z potential	17.52	-26.10	-8.58	-18.19	4.67

IOP Conf. Series: Materials Science and Engineering **121** (2016) 012019 doi:10.1088/1757-899X/121/1/012019

Significant difference was observed between the zeta-potential value of sample site S5 (-26.1mV) and S10 (-13.1 mV). Similar significant difference was observed between S8 (180.3nm) and S10 (386.6nm). For sample site 8, the observed z-potential is -19mV with 180.3 nm as average size which reveals that the sample contains smaller suspended particles with less chance of agglomeration. Its polydispersity index is 0.391 indicating that the sample consists of monodispersed particles [22, 35, 36]. In the sampling location 10, as represented in Figure 2, the observed z-potential is -13.7mV with an average particle size of 386.6 nm signifies the ability of the suspended particles for agglomeration and settlement. From the result, it is evident that the surface lake water samples consist of suspended particles with size smaller than 300nm and with incipient stability. As the surface charge is negative based on z-potential results, chances for heavy metal to be the suspended particles are higher due to the atmospheric deposition and leaching process in the lake [30, 37, 38].

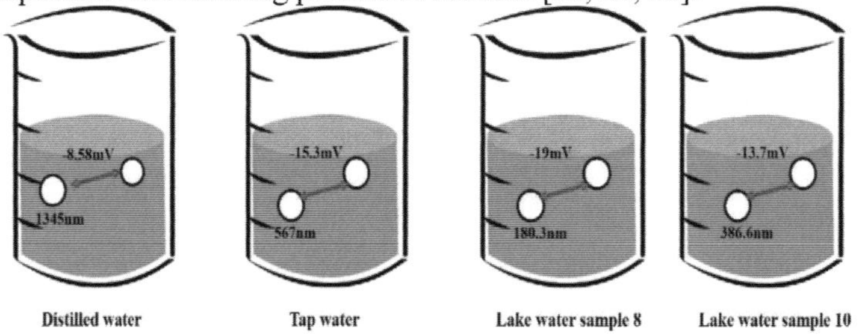

Figure 3. Zeta potential and average size of suspended particles in selected samples.

From the results, we propose that zeta potential can be used for the identification of suspended particle presence in water samples. The particle size distribution along with z-potential will reveal the presence of suspended particles and its ability to agglomerate [39-41]. As the suspended particles are smaller in size, they can easily discharge during the sedimentation and filtration processes which makes them difficult to remove during water treatment [42]. If agglomeration occurs, the size of the particle will increase and eased the filtering process. Figure 3 shows the zeta potential of suspended particles in water samples based on the observed results. From the figure, it is evident that zeta potential plays an important role in the suspended particle analysis and its measurement is believed to be helpful in the water quality monitoring. Moreover, particle size distribution will also contribute in the prediction of larger size particles formation due to agglomeration process.

5. Conclusion

- The presence of suspended particles, its average size and zeta potential was analysed using dynamic light scattering method
- Distilled water was used as reference sample (blank) and the observed results reveals that the higher level of suspended particles in tap water than in drinking water.
- The lake water samples contain smaller suspended particles (lower than 300nm) and their zeta potential discloses that the suspended particles have less chances of agglomeration.
- Thus, it is evident that the suspended particles are smaller in size and free floating in lake water that can affect the aquatic organisms particularly the filter feeding organism.
- Furthermore, we suspect that the negative potential of the suspended particles are due to the presence of heavy metals in the water samples. This can be proved with evidence by heavy metal analysis using atomic absorption spectroscopy which is intended to be performed as future work.

References

[1] Mackas D L 2011 Does blending of chlorophyll data bias temporal trend? *Nature* **472** E4-E5
[2] Rykaczewski R R and Dunne J P 2011 A measured look at ocean chlorophyll trends *Nature* **472** E5-E6
[3] McQuatters-Gollop A, Reid P C, Edwards M, Burkill P H, Castellani C, Batten S, Gieskes W, Beare D, Bidigare R R and Head E 2011 Is there a decline in marine phytoplankton? *Nature* **472** E6-E7
[4] Zinabu G-M 1998 Human interactions and water quality in the Horn of Africa. In: *Science in Africa: Emerging Water Management Issues. Proceedings of a Symposium held in February,*
[5] Prasanna M, Nagarajan R, Chidambaram S and Elayaraja A 2012 Assessment of metals distribution and microbial contamination at selected Lake waters in and around Miri city, East Malaysia *Bulletin of environmental contamination and toxicology* **89** 507-11
[6] Barkman J H and Davidson D H 1972 Measuring water quality and predicting well impairment *Journal of Petroleum Technology* **24** 865-73
[7] Blane J M and Jaakson R 1994 The impact of ecotourism boats on the St Lawrence beluga whales *Environmental conservation* **21** 267-9
[8] Stewart A R, Luoma S N, Schlekat C E, Doblin M A and Hieb K A 2004 Food web pathway determines how selenium affects aquatic ecosystems: a San Francisco Bay case study *Environmental science & technology* **38** 4519-26
[9] Boyle J 2001 *Tracking environmental change using lake sediments*: Springer) pp 83-141
[10] Kelts K and Hsü K 1978 *Lakes*: Springer) pp 295-323
[11] Waters T F 1995 *Sediment in streams: sources, biological effects, and control*: American Fisheries Society)
[12] Lloyd D S, Koenings J P and Laperriere J D 1987 Effects of turbidity in fresh waters of Alaska *North American Journal of Fisheries Management* **7** 18-33
[13] Shinohara R, Imai A, Kawasaki N, Komatsu K, Kohzu A, Miura S, Sano T, Satou T and Tomioka N 2012 Biogenic phosphorus compounds in sediment and suspended particles in a shallow eutrophic lake: a 31P-nuclear magnetic resonance (31P NMR) study *Environmental science & technology* **46** 10572-8
[14] Dawson E and Macklin M 1998 Speciation of heavy metals on suspended sediment under high flow conditions in the River Aire, West Yorkshire, UK *Hydrological Processes* **12** 1483-94
[15] Miller J R 1997 The role of fluvial geomorphic processes in the dispersal of heavy metals from mine sites *Journal of Geochemical Exploration* **58** 101-18
[16] Russell M, Walling D, Webb B and Bearne R 1998 The composition of nutrient fluxes from contrasting UK river basins *Hydrological Processes* **12** 1461-82
[17] Ryan P A 1991 Environmental effects of sediment on New Zealand streams: a review *New Zealand journal of marine and freshwater research* **25** 207-21
[18] Langer O 1980 Effects of sedimentation on salmonid stream life. In: *Report on the Technical Workshop on Suspended Solids and the Aquatic Environment. Whitehorse,*
[19] Cordone A J and Kelley D W 1961 *The influences of inorganic sediment on the aquatic life of streams*: California Department of Fish and Game)
[20] Kemp H A 1949 Soil pollution in the Potomac River basin *Journal (American Water Works Association)* 792-6
[21] Prasanna M, Praveena S, Chidambaram S, Nagarajan R and Elayaraja A 2012 Evaluation of water quality pollution indices for heavy metal contamination monitoring: a case study from Curtin Lake, Miri City, East Malaysia *Environmental Earth Sciences* **67** 1987-2001
[22] Winarti C, Richana N and Sunarti T C 2015 Effect of Acid Hydrolysis on the Characteristics of Andrographolide-Loaded Arrowroot Starch Nanoparticles *International Journal of Food Engineering* **1**

[23] Bean E L, Campbell S J, Anspach F R, Ockershausen R W and Peterman C J 1964 Zeta Potential Measurements in the Control of Coagulation Chemical Doses [with Discussion] *Journal (American Water Works Association)* 214-27
[24] Cohen R D and Probstein R 1986 Colloidal fouling of reverse osmosis membranes *Journal of Colloid and Interface Science* **114** 194-207
[25] Stumm W 1992 Chemistry of the solid-water interface: processes at the mineral-water and particle-water interface in natural systems
[26] Elimelech M, Chen W H and Waypa J J 1994 Measuring the zeta (electrokinetic) potential of reverse osmosis membranes by a streaming potential analyzer *Desalination* **95** 269-86
[27] áO'Brien R W 1990 Electroacoustic studies of moderately concentrated colloidal suspensions *Faraday Discussions of the Chemical Society* **90** 301-12
[28] Hanaor D, Michelazzi M, Leonelli C and Sorrell C C 2012 The effects of carboxylic acids on the aqueous dispersion and electrophoretic deposition of ZrO 2 *Journal of the European Ceramic Society* **32** 235-44
[29] Mateu J, Forteza R, Cerda V and Colom-Altes M 1995 Comparison of various methods for the determination of inorganic species in airborne atmospheric particulates *Water, Air, and Soil Pollution* **84** 61-79
[30] Marsalek R Analysis of Dust Particles in Snow Cover in the Surroundings of the City of Ostrava: Particle Size Distribution, Zeta Potential and Heavy Metal Content *Analysis* **1** 12754
[31] Wilson W W, Wade M M, Holman S C and Champlin F R 2001 Status of methods for assessing bacterial cell surface charge properties based on zeta potential measurements *Journal of Microbiological Methods* **43** 153-64
[32] Sprycha R 1989 Electrical double layer at alumina/electrolyte interface: I. Surface charge and zeta potential *Journal of colloid and interface science* **127** 1-11
[33] Fengqiu T, Xiaoxian H, Yufeng Z and Jingkun G 2000 Effect of dispersants on surface chemical properties of nano-zirconia suspensions *Ceramics International* **26** 93-7
[34] Li D, Mueller M B, Gilje S, Kaner R B and Wallace G G 2008 Processable aqueous dispersions of graphene nanosheets *Nature nanotechnology* **3** 101-5
[35] de Pinho Neves A L, Milioli C C, Müller L, Riella H G, Kuhnen N C and Stulzer H K 2014 Factorial design as tool in chitosan nanoparticles development by ionic gelation technique *Colloids and Surfaces A: Physicochemical and Engineering Aspects* **445** 34-9
[36] Zhang L and Kosaraju S L 2007 Biopolymeric delivery system for controlled release of polyphenolic antioxidants *European Polymer Journal* **43** 2956-66
[37] Appel C and Ma L 2002 Concentration, pH, and surface charge effects on cadmium and lead sorption in three tropical soils *Journal of Environmental Quality* **31** 581-9
[38] Abbracchio M P, Heck J D and Costa M 1982 The phagocytosis and transforming activity of crystalline metal sulfide particles are related to their negative surface charge *Carcinogenesis* **3** 175-80
[39] O'brien R, Cannon D and Rowlands W 1995 Electroacoustic determination of particle size and zeta potential *Journal of Colloid and Interface Science* **173** 406-18
[40] Wiese G and Healy T 1970 Effect of particle size on colloid stability *Transactions of the Faraday Society* **66** 490-9
[41] Jiang J, Oberdörster G and Biswas P 2009 Characterization of size, surface charge, and agglomeration state of nanoparticle dispersions for toxicological studies *Journal of Nanoparticle Research* **11** 77-89
[42] Incorporation Z M 2005 Zeta Potential: A Complete Course in 5 Minutes *Zeta-Meter, Inc* **765**

Acknowledgment
The authors are greatly indebted to Curtin University, Sarawak campus, Malaysia for providing the necessary facilities to carry out the research.

IOP Conf. Series: Materials Science and Engineering **121** (2016) 012020 doi:10.1088/1757-899X/121/1/012020

Bi_2O_3 and $La_{10}Si_6O_{27}$ composite electrolyte for enhanced performance in solid oxide fuel cells

H Q Hj Hairul Absah[1, 3], M S Abu Bakar[1], J Hj Zaini[1], A Azad[1] and L C Ming[2]
[1]Universiti Brunei Darussalam, Faculty of Integrated Technologies, Jalan Tungku Link, Gadong BE 1410, Brunei Darussalam
[2]Universiti Brunei Darussalam, Centre for Advanced Material and Energy Sciences, Jalan Tungku Link, Gadong BE 1410, Brunei Darussalam
[3]Corresponding author
E-mail: 13H8058@ubd.edu.bn, atulsimah@gmail.com

Abstract. Adding suitable metal oxide into lanthanum silicate apatite can produce a composite with a good oxygen ion-conducting electrolyte that enhances the performance of solid oxide fuel cells (SOFCs). In this paper we present the synthesis and characterisation of Bi_2O_3 and $La_{10}Si_6O_{27}$ composite prepared by a solid state reaction. The sintering temperature of the composite was 1500°C for 10 hours with the heating and cooling rates of 10°C per minute. The properties of the resulting composite have been characterised by X-ray diffraction (XRD), scanning electron microscopy (SEM), and ionic conductivity measured by an a.c. impedance spectroscopy (IS). Rietveld refinement of XRD data shows that the composition is purely the mixture of Bi_2O_3 and $La_{10}Si_6O_{27}$ with the unit cell parameters of the main phase as a = 9.9810 (8) and c = 7.3239 (6) Å. The room temperature crystal structure was hexagonal with space group $P6_3/m$. The highest ionic conductivity of 1.76 x 10^{-2} Scm^{-1} with a corresponding activation energy of 0.39 eV was obtained at 750°C. SEM images show the material is densed enough to use as an electrolyte for SOFCs.

1. Introduction

Solid oxide fuel cell (SOFC) is a type of electrochemical device that receives much research attention over the years due to its sustainable energy systems, highly efficient conversion of chemical energy into electrical energy, fuel flexibility and environmentally cleans power generation [1-3]. SOFC systems consist of a dense electrolyte that separates the anode and the cathode in a unit SOFC. Low to intermediate operating temperatures SOFC (400 to 800°C) give significant advantages such as wider range of material selection, longer life and lower manufacturing costs. Many intensive researches have been done to produce solid oxide electrolytes that exhibit high oxide ionic conductivity even at the intermediate temperatures (600 to 800°C) to replace yttria-stabilized zirconia (YSZ) that operates at temperatures as high as 900 to 1000°C [4-8].

Doped ceria and lanthanum gallate perovskite are among the solid oxide electrolytes that exhibit higher oxide ion conductivity at the intermediate operating temperatures. However, chemical instability and degradation of GDC10, LSGM and Y_2O_3-Bi_2O_3 electrolytes have led to several research efforts for alternative, low cost and stable solid oxide electrolytes that can produce high oxide ion conductivity at relatively low to intermediate operating temperatures [9].

Recently some new high oxide ion conductivity solid oxide electrolytes at intermediate operating temperature have been produced to overcome these problems. Among the list was the apatite based on rare-earth silicates with a chemical formula $RE_{10}Si_6O_{27}$ (RE = La, Pr, Nd, Sm, Gd and Dy) [10]. $La_{10}Si_6O_{27}$ apatite has been reported as having the highest oxide ion conductivity compare to other apatite type rare earth silicates [11-16]. Thus $La_{10}Si_6O_{27}$ apatite attracts most of the attention. Other advantages of $La_{10}Si_6O_{27}$ apatite are purely ionic conductivity with oxygen transference numbers near unity over a wide oxygen partial pressure range and stable electrochemical performance under various gases feed stocks [11, 17-20].

In recent years, many low cost production methods have been applied to eliminate secondary products of $La_{10}Si_6O_{27}$ apatite (La_2SiO_5 and La_2O_3), reduce the complete formation temperature of $La_{10}Si_6O_{27}$ apatite and produce dense $La_{10}Si_6O_{27}$ apatite with high oxide ion conduction [21-25]. Benefits of various dopants have also been studied to improve the ionic conductivity of $La_{10}Si_6O_{27}$ apatite that increases the interstitial oxide ions concentration in the electrolytes. Two potential atomic sites for doping $La_{10}Si_6O_{27}$ apatite are the La and Si sites. Si site doping has shown to increase the overall ionic conductivity of $La_{10}Si_6O_{27}$ apatite than the La site [11, 26]. Most recent research has shown that doping metal oxide into $La_{10}Si_6O_{27}$ apatite resulted in a dense solid oxide electrolyte at low sintering temperature and high oxide ion conductivity at the intermediate operating temperature [27]. This indicates that doping $La_{10}Si_6O_{27}$ apatite with metal oxide can improve the performance of the $La_{10}Si_6O_{27}$ apatite.

In this study, solid-state reaction was used to obtain Bi_2O_3-$La_{10}Si_6O_{27}$ composite electrolyte. The objectives of this study are to obtain pure Bi_2O_3-$La_{10}Si_6O_{27}$ composite electrolyte under low sintering temperature and investigate the oxide ion conductivity at low to intermediate operating temperatures.

2. Experimental

2.1. Sample preparation and characterisation

The metal oxide powders of 5 weight % Bi_2O_3 (China) and 95 weight % $La_{10}Si_6O_{27}$ (77.8 weight % La_2O_3 (Merck) and 17.2 weight % SiO_2 (Fluka Analytical Sigma-Aldrich)) were used as an initial materials. Appropriate amounts of the powders were mixed with distilled water. Two drops of polyvinyl alcohol was added to the mixture as a binder. The ceramic slurry was then milled with alumina balls using ABBE Jar Rolling Mills (model: 202241) for 24 hours and dried completely in an oven at a temperature of 100°C. The ceramic powder obtained after drying the slurry was ground and 1 g of the powder was pressed uniaxially using a hydraulic pressing device in a 1.3 cm diameter mold at a constant pressure of 4 bars and a hold up time of 3 minutes. The prepared pellets were sintered at 1500°C for 10 hours in the presence of air with the heating and cooling rates of 10°C/min to obtain dense ceramic composites. Room temperature X-ray diffraction pattern of the ceramic composite was obtained on SHIMADZU XRD-6000/7000 (version: 5.21) using Cu-Kα1 radiation (wavelength λ = 1.5406 Å) at 40 kV and 30 mA. A sampling pitch of 0.02 degree and a scan speed of 2 degree per minute were used to determine the purity and crystal parameters. Microstructural analysis was carried out on a JEOL JSM-7610F scanning electron microscopy. The pellet densities ρ were measured using the relation

$$\rho = \frac{mass\ of\ pellet}{volume\ of\ pellet} \quad (1)$$

The pellet densities were compared with the theoretical density ρ_{th} obtained using the relation

$$\rho_{\text{th}} = \frac{\textit{molecular weight of compound}}{\mathrm{N_A}.\frac{\sqrt{3}}{2}.a^2.c} \quad (2)$$

where $\mathrm{N_A}$ is the Avogadro constant (6.022 x 10^{23} mol^{-1}), a, and c are the lattice parameters of the ceramic composite (cm).

2.2. Conductivity Measurements

Conductivity measurements were performed using an in-house designed furnace with two platinum wires used as current collectors. A symmetrical cell with silver paste coating on each side of the pellet was made. The coated pellet was heated at 700°C for 2 hours to burn any organic additives used in the paste. A.C. impedance measurements were collected in 50°C steps in air between 400 and 750°C using a Solartron impedance analyzer system (Solartron™ SI 1260, UK) combined with electrochemical interface (Solartron™ SI 1287, UK) controlled by Zplot electrochemical impedance software. An excitation voltage of 50 mV was used between 0.1 and 1 MHz. Total resistance (sum of bulk and grain boundary resistances) was obtained from fitting the impedance plot. The total conductivity σ_{total} at different temperatures was evaluated using the following equation:

$$\sigma_{\text{total}} = \frac{t}{SR} \quad (3)$$

where t is the thickness of the pellet, S is the surface area of the pellet and R is the total resistance.

Activation energy was then obtained from Arrhenius plot using Arrhenius equation:

$$\sigma T = \sigma_o \exp\left(\frac{E_a}{kT}\right) \quad (4)$$

where $\sigma, \sigma_o, E_a, \mathrm{k}, T$ are the conductivity (Scm^{-1}), pre-exponential factor, activation energy (eV), Boltzmann constant (8.62 x 10^{-5} eV/K) and temperature (K), respectively. After a simple arrangement, equation (4) can be written as:

$$\log(\sigma T) = -\frac{E_a}{\mathrm{k}}.\frac{1}{T} + \log(\sigma_o) \quad (5)$$

3. Results and discussions

Effect of adding bismuth oxide, milling and sintering of the ceramic composite were analysed.

3.1. Phase identification

Room temperature X-ray diffraction pattern of Bi_2O_3-$La_{10}Si_6O_{27}$ composite sintered at 1500°C is shown in figure 1. The main XRD pattern of Bi_2O_3-$La_{10}Si_6O_{27}$ is identical with the standard XRD spectrum of $La_{10}Si_6O_{27}$ (JCPDS File No. 53-0291). Rietveld refinement of XRD data using the Fullprof suite [28] confirmed a purely single phase Bi_2O_3-$La_{10}Si_6O_{27}$ belonging to the hexagonal apatite structure with space group $P6_3/m$ and the lattice parameter of the main phase as a = 9.9810 (8) and c = 7.3239 (6) Å. The bulk densities of the pellets after sintering at 1500°C range from 2.8 to 3.4 gcm^{-3} (55.7 to 67.6% of theoretical density). The theoretical density of the pellet is 5.028 gcm^{-3}.

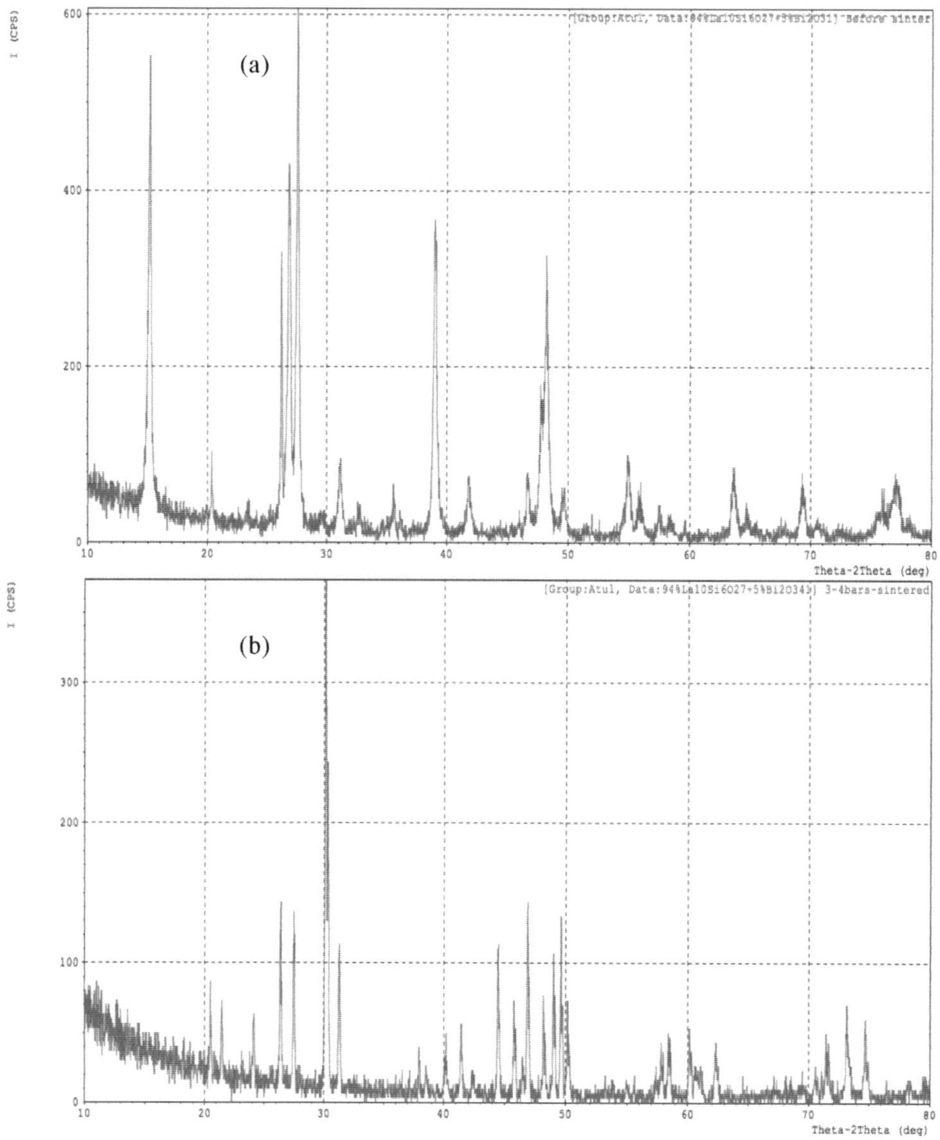

Figure 1. (a) Room temperature XRD of Bi_2O_3-$La_{10}Si_6O_{27}$ composite before sintering and (b) Room temperature XRD of Bi_2O_3-$La_{10}Si_6O_{27}$ composite after sintering at 1500°C.

3.2. Microstructural analysis

Figure 2 presents the microstructure of Bi_2O_3-$La_{10}Si_6O_{27}$ composite. SEM micrographs of the ceramic composite show compact hexagonal apatite structure with average particle size from 1 to 10 μm. Bi_2O_3-$La_{10}Si_6O_{27}$ composite is also quite dense with a low porosity at the low magnification.

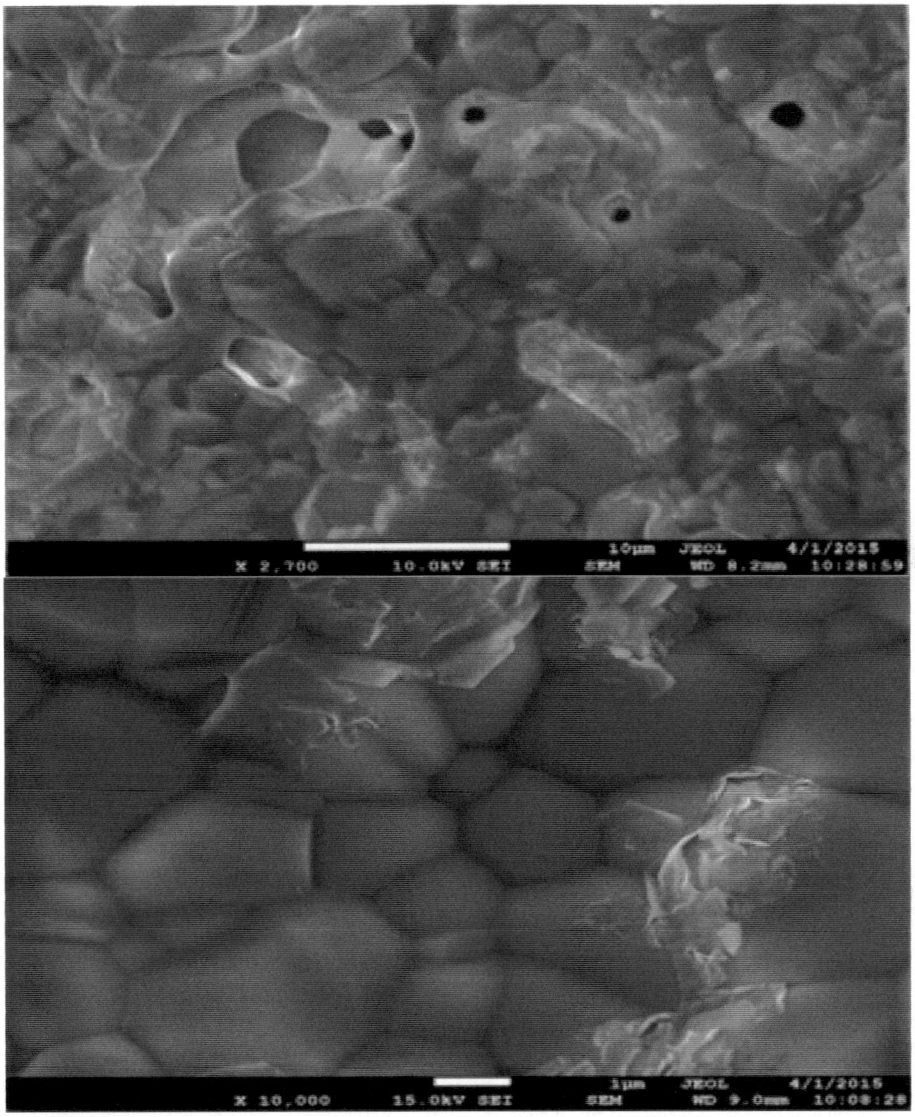

Figure 2. SEM micrographs of Bi_2O_3-$La_{10}Si_6O_{27}$ composite sintered at 1500°C.

3.3. Conductivity measurements

Figure 3 shows the complex impedance plots of sintered Bi_2O_3-$La_{10}Si_6O_{27}$ composite at different temperatures from 400 to 750°C. The relative density (measured density over theoretical density) of this pellet is 67.6%. The impedance plots at 400 to 550°C exhibit two semicircles that present grain and grain boundary arcs. However, the grain and grain boundary arcs combine with increasing temperature from 600 to 750°C resulting only one semicircular arc. As a consequence, the total resistance value was used to determine the total conductivity value at different temperatures. The corresponding equivalent circuits for different impedance plots are also shown in figure 3.

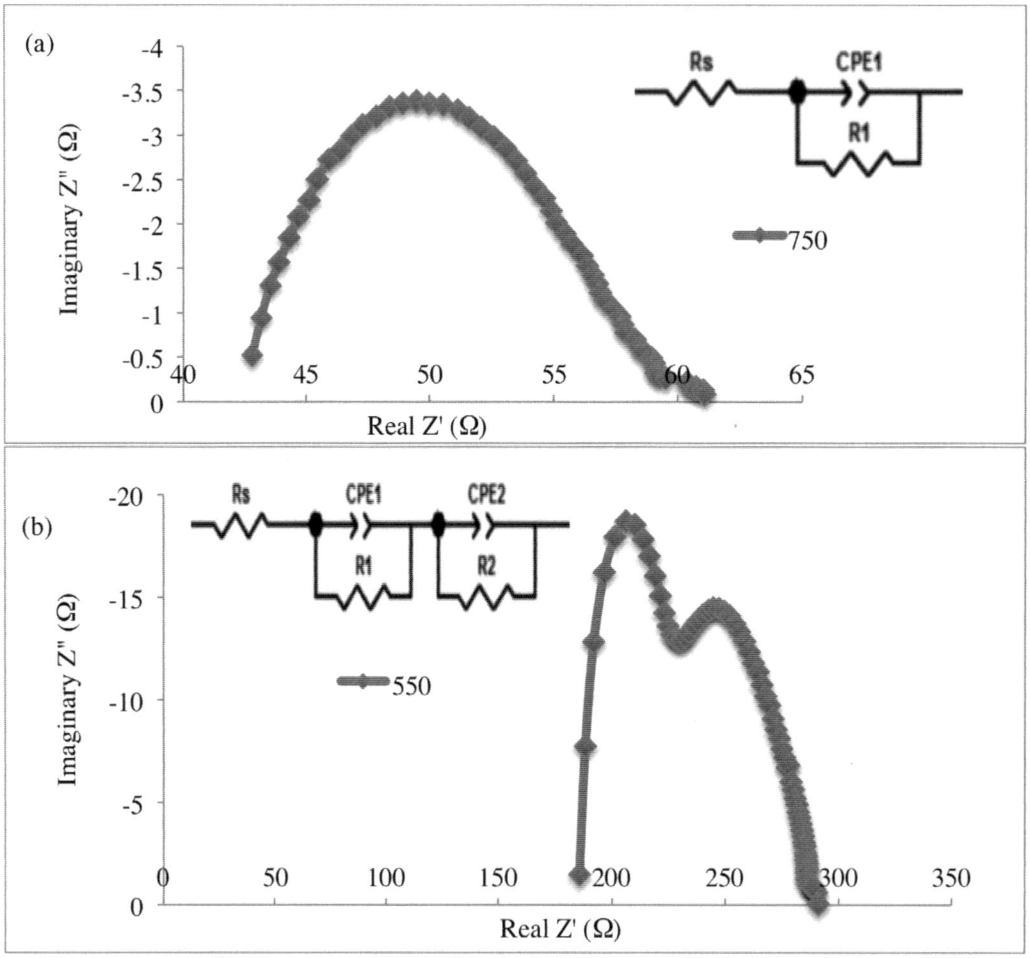

Figure 3. A.C. impedance plots of Bi_2O_3-$La_{10}Si_6O_{27}$ composite and equivalent circuits at (a) 750°C and (b) 550°C.

The corresponding bulk resistance R_1, grain boundary resistance R_2 and total resistance R_{total} from fitting results are tabulated in table 1. The total conductivity of Bi_2O_3-$La_{10}Si_6O_{27}$ composite at different temperatures is calculated using equation (3). The total conductivity σ_{total} of the ceramic composite gradually increases with increasing temperature from 400 to 750°C as shown in table 2, which demonstrates that the ionic diffusion process is thermally activated.

Table 1. Resistance of bismuth oxide-lanthanum silicate composite from fitting results.

Temperature (°C)	R_1 (Ω)	R_2 (Ω)	R_{total} (Ω)
400	1565	725	2290
450	450	287	737
500	115	150	265
550	30	77	107
600	51.5	-	51.5
650	34.2	-	34.2
700	34	-	34
750	16	-	16

Table 2. Conductivity of bismuth oxide-lanthanum silicate composite.

Temperature (°C)	σ_{bulk} (Scm^{-1})	$\sigma_{grain\ boundary}$ (Scm^{-1})	σ_{total} (Scm^{-1})
400	1.80 x 10^{-4}	3.89 x 10^{-4}	1.23 x 10^{-4}
450	6.27 x 10^{-4}	9.84 x 10^{-4}	3.83 x 10^{-4}
500	2.46 x 10^{-3}	1.88 x 10^{-3}	1.07 x 10^{-3}
550	9.41 x 10^{-3}	3.67 x 10^{-3}	2.64 x 10^{-3}
600	-	-	5.48 x 10^{-3}
650	-	-	8.26 x 10^{-3}
700	-	-	8.3 x 10^{-3}
750	-	-	1.76 x 10^{-2}

From table 2, the bulk conductivity σ_{bulk} and grain boundary conductivity $\sigma_{\text{grain boundary}}$ increase from 400 to 550°C. Figure 4 presents the Arrhenius plot of the ceramic composite. Straight line can be drawn from the Arrhenius plot that is well fitted to the Arrhenius equation. The fitted line also indicates that the diffusion process of oxide ions is thermally activated. From the slope and the intercept of the linear fits in the Arrhenius plot, the activation energy E_a and pre-exponential factor σ_o of the ceramic composite can be obtained using equation (5). The values of activation energy E_a and pre-exponential factor σ_o are 0.39 eV and 5.21 x 10^5 SKcm^{-1} respectively. From table 2, the highest total conductivity of Bi_2O_3-$La_{10}Si_6O_{27}$ composite is 1.76 x 10^{-2} Scm^{-1} at 750°C. The addition of 5 weight % Bi_2O_3 into 95 weight % $La_{10}Si_6O_{27}$ resulted in a good conductivity achieved at 750°C for sintering at 1500°C in the presence of air. Unfortunately, dense Bi_2O_3-$La_{10}Si_6O_{27}$ pellets could not be obtained for conductivity measurements because the pellets are not fully sintered even after heating at the maximum operating temperature (1500°C) of the furnace. Bi_2O_3-$La_{10}Si_6O_{27}$ composite exhibits a significant enhancement on the total conductivity at 750°C for sintering at 1500°C in air. This makes the composite electrolyte suitable for intermediate temperature solid oxide fuel cell.

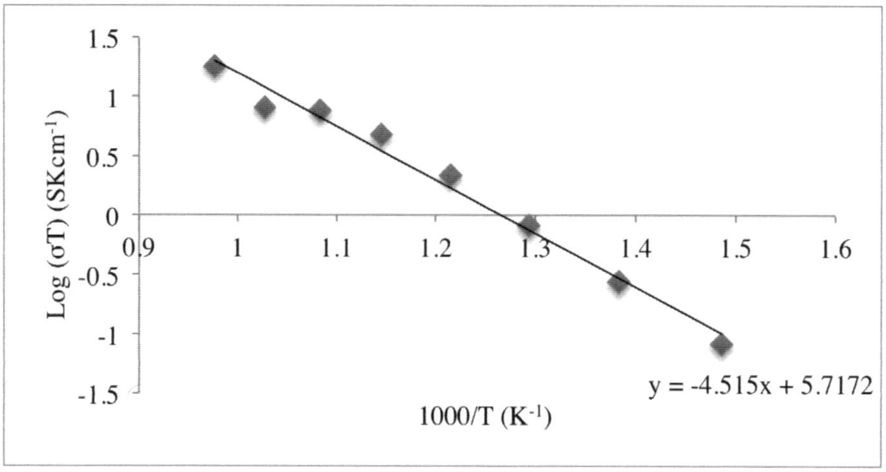

Figure 4. Arrhenius plot of the total conductivity of Bi_2O_3-$La_{10}Si_6O_{27}$ composite.

4. Conclusion

Pure apatite-type Bi_2O_3-$La_{10}Si_6O_{27}$ composite electrolyte has been successfully prepared by the solid-state reaction. The room temperature structure of the ceramic composite is hexagonal, space group $P6_3/m$, with a = 9.9810 (8) and c = 7.3239 (6) Å. The total conductivity obeys the Arrhenius relation, and gradually increases with increasing temperature from 400 to 750°C. The highest total conductivity is 1.76 x 10^{-2} Scm^{-1} at 750°C. This experiment concludes that Bi_2O_3 has a potential to produce pure Bi_2O_3-$La_{10}Si_6O_{27}$ ceramic composite electrolyte at low sintering temperature and improve the conductivity of $La_{10}Si_6O_{27}$ at low to intermediate operating temperatures.

5. References

[1] Brandon N P, Skinner S and Steele B C H 2003 Recent advances in materials for fuel cells *Annual Rev. Mater. Res.* **33(1)** 183–213
[2] Dokiya M 2002 SOFC system and technology *Sol. State Ion.* **152-3** 383–92
[3] Sharaf O Z and Orhan M F 2014 An overview of fuel cell technology: fundamentals and applications *Renew. and Sustain. Energy Rev.* **32** 810–53
[4] Azad A K and Irvine J T S 2007 Synthesis, chemical stability and proton conductivity of the perovksites Ba(Ce,Zr)1−x Scx O3−δ *Sol. State Ion.* **178(7–10)** 635–40
[5] Ghosh A, Azad A and Irvine J T 2011 Study of Ga doped LSCM as an anode for SOFC *ECS Trans.* **35(1)** 1337-43
[6] Zhang X, Chan S H, Li G, Ho H K, Li J and Feng Z 2010 A review of integration strategies for solid oxide fuel cells *J. Pow. Sour.* **195(3)** 685–702
[7] Fergus J W 2006 Electrolytes for solid oxide fuel cells *J. Pow. Sour.* **162(1)** 30–40
[8] Jacobson A J 2010 Materials for solid oxide fuel cells *Chem. Mater.* **22(3)** 660–74
[9] Mogensen M, Sammes N M and Tompsett G A 2000 Physical, chemical and electrochemical properties of pure and doped ceria *Sol. State Ion.* **129(1-4)** 63–94
[10] Nakayama S and Sakamoto M 1998 Electrical properties of new type high oxide ionic conductor RE10Si6O27 (RE = La, Pr, Nd, Sm, Gd, Dy) *J. Euro. Ceram. Soc.* **18(10)** 1413–18
[11] Marrero-López D, Santos-Gómez L D, León-Reina L, Canales-Vázquez J and Losilla E R 2014 Influence of the microstructure on the bulk and grain boundary conductivity in apatite-type electrolytes *J. Pow. Sour.* **245** 107–18

[12] Marrero-López D, Martín-Sedeño M C, Peña-Martínez J, Ruiz-Morales J C, Núñez P, Aranda M A G and Ramos-Barrado J R 2010 Evaluation of apatite silicates as solid oxide fuel cell electrolytes *J. Pow. Sour.* **195(9)** 2496–2506
[13] Liu W, Yamaguchi S, Tsuchiya T, Miyoshi S, Kobayashi K and Pan W 2013 Sol–gel synthesis and ionic conductivity of oxyapatite-type La9.33+xSi6O26+1.5x *J. Pow. Sour.* **235** 62–6
[14] Zhang Y, Su Z, Azad A K, Zhou W and Irvine J T S 2012 Direct imaging interstitial oxygen in silicate apatite *Advanced Energy Mat.* **2(3)** 316-21
[15] Porras-Vázquez J M, Losilla E R, Aranda M A G and Santacruz I 2012 Oxy-apatite reaction sintering of colloidal and classic ceramic processed powders *Ceram. Int.* **38(3)** 1851–58
[16] Rodríguez-Reyna E, Fuentes A F, Maczka M, Hanuza J, Boulahya K and Amador U 2006 Facile synthesis, characterization and electrical properties of apatite-type lanthanum germanates *Sol. State Sci.* **8(2)** 168–177
[17] Shaula A L, Kharton V V, Waerenborgh J C, Rojas D P and Marques F M B 2005 Oxygen ionic and electronic transport in apatite ceramics *J. Sol. State Chem.* **178(6)** 2050–61
[18] Cao X G and Jiang S P 2012 Sinterability and conductivity of barium doped aluminium lanthanum oxyapatite La9.5Ba0.5Si5.5Al0.5O26.5 electrolyte of solid oxide fuel cells *J. Alloys and Comp.* **523** 127–133
[19] Nakao T, Mineshige A, Kobune M, Yazawa T and Yoshioka H 2008 Chemical stability of La10Si6O27 and its application to electrolytes for solid oxide fuel cells *Sol. State Ion.* **179(27-32)** 1567–69
[20] Marrero-López D, Peña-Martínez J, Ruiz-Morales J C, Gabás M, Núñez P, Aranda M A G and Ramos-Barrado J R 2010 Redox behaviour, chemical compatibility and electrochemical performance of Sr2MgMoO6 - δ as SOFC anode *Sol. State Ion* **180(40)** 1672–82
[21] Tao S and Irvine J T S 2001 Preparation and characterisation of apatite-type lanthanum silicates by a sol-gel process *Mat. Res. Bull.* **36(7-8)** 1245–58
[22] Yamagata C, Elias D R, Paiva M R S, Misso A M and Castanho S R H M 2013 Facile preparation of apatite-type lanthanum silicate by a new water-based sol-gel process *Mat. Res. Bull.* **48(6)** 2227–31
[23] Jothinathan E, Vanmeensel K, Vleugels J and Van der Biest O 2010 Synthesis of nano-crystalline apatite type electrolyte powders for solid oxide fuel cells *J. Euro. Ceram. Soc* **30(7)** 1699–1706
[24] Shi Q, Lu L, Jin H, Zhang H and Zeng Y 2012 Electrical properties and thermal expansion of cobalt doped apatite-type lanthanum silicates based electrolytes for IT-SOFC *Mat. Res. Bull.* **47(3)** 719–23
[25] Tian C, Liu J, Cai J and Zeng Y 2008 Direct synthesis of La9.33Si6O26 ultrafine powder via sol-gel self-combustion method *J. Alloys and Comp.* **458(5)** 378–82
[26] Xiang J, Liu Z G, Ouyang J H and Yan F Y 2014 Ionic conductivity of oxy-apatite La10Si6–xInxO27–δ solid electrolyte ceramics *J. Pow. Sour.* **251** 305–10
[27] Setsoafia D D Y, Hing P, Jung S C, Azad A K and Lim C M 2015 Sol–gel synthesis and characterization of Zn2+ and Mg2+ doped La10Si6O27 electrolytes for solid oxide fuel cells *Sol. State Sci.* **48** 163–170
[28] Carvajal J R 1993 Recent advances in magnetic structure determination by neutron powder diffraction *Physica* B **192(1-2)** 55-69

Acknowledgment

Author, H Q Hj Hairul Absah thanks the Ministry of Education, Brunei Darussalam for giving scholarship to conduct this research at the Universiti Brunei Darussalam. Authors also thank Universiti Brunei Darussalam Research Grant UBD/PNC2/2/RG/1(176), Brunei Research Council (BRC) Science and Technology grant S&T 17, Faculty of Integrated Technologies, Faculty of Science and Centre for Advanced Material and Energy Sciences for the support.

A unified double-loop multi-scale control strategy for NMP integrating-unstable systems

Qiu Han Seer[1,2] and Jobrun Nandong[2,3]

[1] Curtin Sarawak Research Institute, Curtin University Sarawak, 98009 Miri, Sarawak, Malaysia

[2] Department of Chemical Engineering, Curtin University Sarawak, 98009 Miri, Sarawak, Malaysia

E-mail: jobrun.n@curtin.edu.my

Abstract. This paper presents a new control strategy which unifies the direct and indirect multi-scale control schemes via a double-loop control structure. This unified control strategy is proposed for controlling a class of highly nonminimum-phase processes having both integrating and unstable modes. This type of systems is often encountered in fed-batch fermentation processes which are very difficult to stabilize via most of the existing well-established control strategies. A systematic design procedure is provided where its applicability is demonstrated via a numerical example.

1. Introduction

Proportional-integral-derivative (PID) controller has widely been used in process control industries due to its simple structure, robust performance and ease in implementation in practice [1–3]. The basic idea of PID controller is depended on the tuning of controller parameters controller gain K_c, integral time τ_I and τ_D, sometimes with filter time constant τ_f in order to obtain the accuracy and performance of controllers.

The unstable and integrating systems are usually observed in process industries, such as batch reactors in the fermentation industry [4]. In bioprocesses, the strong nonlinearity characteristics are frequently caused by process variability and complexity of biological systems [5]. Other than fermentation process, the unstable and integrating process are frequently occurred in chemical plant industries related to heat generation. The unstable processes are difficult to control compared to open-loop stable process due to the difficulty for stabilizing unstable poles, which can cause the instability of system and lead to poor control performance. However, the conventional PID controllers are normally employed for stable processes [1], which are difficult to design for integrating and unstable processes [3].

Due to the limitations imposed by integrating and unstable processes, there are different tuning approaches which have been introduced by many researchers. [4, 6] proposed an IMC-based PID tuning rules for open-loop unstable process models without time delay, whilst [7] introduced a similar tuning method for delayed unstable process. [8] developed a graphical tuning method for integrating and unstable processes with time delay based on the analysis of the open-loop frequency response of the process on the Nichols chart. [9] suggested an integrated

[3] Corresponding author

design procedure for a modified Smith predictor and associated controller for unstable processes with time delay. [10] derived a tuning method for unstable processes with time delay based on gain and phase margins for P, PI and PID controllers. Also, [11] developed a simple fuzzy PID controller tuning method while [12] proposed a self-tuning fuzzy PID controller for integrating process with dead time. [1,3] implemented an additional inner feedback loop in order to stabilize the integrating and unstable systems.

This paper presents an idea of incorporating an inner feedback loop with the multi-scale control (MSC) scheme. The MSC scheme offers a systematic approach for designing multi-loop PID controllers augemented with filters, which provide enhanced performance robustness [13]. Meanwhile, the inner feedback loop is able to stabilize the unstable mode or pole, i.e. using a simple P controller, before the implementation of MSC-PID controller in the outer loop.

In Section 2, a brief overview of MSC scheme is introduced. Section 3 presents the derivation of MSC-PID tuning relations. In Section 4, the control system design is presented, which includes the details of tuning method of secondary loop in combination with the MSC scheme. Also, a general procedure of designing the proposed control system is provided. In Section 5, an illustrative example based on a fed-batch bioethanol production is used to demonstrate the effectiveness of the proposed control scheme. At last, Section 6 highlights some conclusions and recommendations for future research directions.

2. Preliminaries

2.1. Multi-Scale Control Scheme

The details of multi-scale control (MSC) scheme can be found in [14–17]. A brief overview of the MSC scheme is presented. Fig. 1 represents the block diagram of a 2-layer direct multi-scale control (MSC) scheme while the corresponding indirect MSC scheme is shown in Fig. 2, where the plant P can be decomposed into a sum of two basic modes as follows:

$$P(s) = m_0(s) + m_1(s) \tag{1}$$

where m_0 represent the outermost (slow) mode and m_1 represent the inner-layer (fast) mode, which can be first or second order system with real coefficients. For the direct MSC scheme, the multi-scale predictor is often chosen as the faster inner mode. On the other hand, the multi-scale predictor is chosen as the slower outermost mode for the indirect MSC scheme.

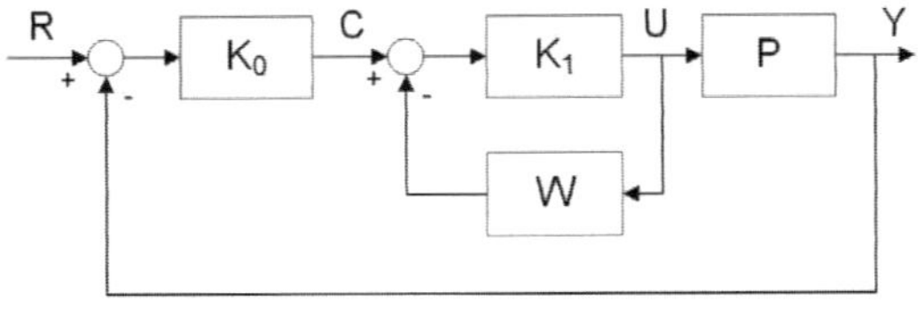

Figure 1. Direct MSC scheme

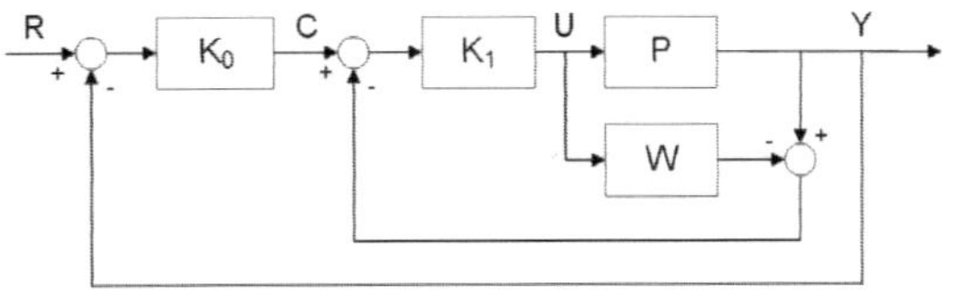

Figure 2. Indirect MSC scheme

In Fig. 1 and Fig. 2, K_i, $i = 0, 1$ represent the sub-controller of outermost and inner-layer modes, W is the multi-scale predictor, R and Y denote the setpoint and controlled variable signals respectively. The closed-loop transfer function for the inner-layer is defined as

$$G_1(s) = \frac{U(s)}{C(s)} = \frac{K_1(s)}{1 + K_1(s)W(s)} \tag{2}$$

The two-layer MSC scheme can now be reduced to single-loop feedback control and the overall MSC controller can be defined as follows:

$$G_c(s) = K_0(s)G_1(s) \tag{3}$$

3. Derivation of MSC-PID Tuning Relations

A second order integrating-unstable process is considered after partial stabilization in order to present a brief construction procedure of MSC-PID tuning formula.

$$P(s) = \frac{K_p(-\tau_z s + 1)}{s(\tau_p s + 1)} \tag{4}$$

where K_p, τ_z and τ_p represent the process gain, lead time constant and time constant respectively.

The partial fraction expansion is applied to decompose (4) into a sum of two basic modes given by

$$P(s) = \underbrace{\frac{k_{p0}}{s}}_{m_0} + \underbrace{\frac{k_{p1}}{\tau_p s + 1}}_{m_1} \tag{5}$$

where m_0 and m_1 denote the outermost and inner-layer modes, respectively. The mode gains are given by

$$k_{p0} = K_p \tag{6}$$

$$k_{p1} = -K_p(\tau_z + \tau_p) \tag{7}$$

Assume that a P-only controller is used for controlling the inner-layer mode m_1 and a PI controller is used for controlling the outermost mode m_0. The sub-controllers for inner-layer and outermost modes are given by

$$K_1 = k_{c1} \tag{8}$$

$$K_0 = k_{c0}(1 + \frac{1}{\tau_{I0} s}) \tag{9}$$

where k_{c1} and k_{c0} denote the sub-controller gains of inner-layer and outermost mode respectively; τ_{I0} is the integral time for the outermost mode sub-controller.

Based on the direct MSC scheme (refer to Fig. 1), assume that the multi-scale predictor is chosen as the fast inner mode, the inner-layer transfer function is defined as

$$G_1(s) = \frac{k_{c1}}{1 + k_{c1}W(s)} \tag{10}$$

After substituting the inner mode $W(s) = \frac{k_{p1}}{\tau_p s+1}$ into (10) and followed by simplification

$$G_1(s) = \frac{k_{c1}^0(\tau_p s + 1)}{\tau_{c1} s + 1} \tag{11}$$

where the overall gain and closed-loop time constant are given in term of

$$k_{c1}^0 = \frac{k_{c1}}{1 + k_{c1}k_{p1}} \tag{12}$$

$$\tau_{c1} = \frac{\tau_p}{1 + k_{c1}k_{p1}} \tag{13}$$

IOP Conf. Series: Materials Science and Engineering **121** (2016) 012021 doi:10.1088/1757-899X/121/1/012021

The ratio of open loop time constant to closed-loop time constant is defined as follows:

$$\lambda_1 = \frac{\tau_p}{\tau_{c1}}; \quad \lambda_1 > 1 \tag{14}$$

From (13) and (14), k_{c1} can be expressed as follows:

$$k_{c1} = \frac{\lambda_1 - 1}{k_{p1}}; \quad \lambda_1 > 1 \tag{15}$$

Meanwhile, PI controller is chosen to control the outermost mode. However, P-only controller is assumed first in order to determine the k_{c0} following the same way as the inner-layer mode.

$$k_{c0} = \frac{\lambda_0 - 1}{k_{p0}} \tag{16}$$

However, the open-loop time constant is undefined for an integrating mode. Thus, the range of λ_0 is unclear.

In order to calculate k_{c0}, P-only controller with gain k_{c0} is used based on the unity feedback control and the characteristic equation by using (4) is given by

$$\tau_p s^2 + (1 - k_{c0}K_p\tau_z)s + k_{c0}K_p = 0 \tag{17}$$

Based on the Routh Stability criterion, the upper limits can be written in term of a parameter r_1 as follows:

$$k_{c0}K_p < \frac{1}{\tau_z} = \frac{1}{r_1\tau_z}; \quad r_1 > 1 \tag{18}$$

From (16) and (18), the k_{c0} can be calculated by applying (6) that $k_{p0} = K_p$ in the following manner.

$$k_{c0} = \frac{1}{r_1\tau_z K_p} = \frac{\lambda_0 - 1}{K_p}; \quad r_1 > 1 \tag{19}$$

Thus, the outermost time constant λ_0 can be obtained from (19) as follows:

$$\lambda_0 = \frac{r_1\tau_z + 1}{r_1\tau_z}; \quad r_1 > 1 \tag{20}$$

Thus, $\lambda_0 > 1$ in order to stabilize the outermost mode.

Let us define an equivalent ('fictitious') open-loop time constant τ_∞ for the integrating mode, similar to (14).

$$\lambda_0 = \frac{\tau_\infty}{\tau_{c0}}; \quad \lambda_0 > 1 \tag{21}$$

Note that, τ_{c0} represents the closed-loop time constant for the outermost mode as follows:

$$\tau_{c0} = \frac{1}{k_{c0}K_p} = r_1\tau_z \tag{22}$$

Substitute (20) and (22) into (21) in order to get the open-loop time constant τ_∞.

$$\tau_\infty = r_1\tau_z + 1 \tag{23}$$

The integral time for the outermost mode is expressed based on desired fraction γ of the open-loop time constant τ_∞.

$$\tau_{I0} = \gamma \tau_{\infty} \tag{24}$$

where a range of $0.5 \leq \gamma \leq 3.0$ is recommended.

The overall MSC controller for the partially stabilized plant can be arranged as a PID controller with filter in the classical PID form.

$$G_c(s) = K_c(1 + \frac{1}{\tau_I s})(\frac{\tau_D s + 1}{\tau_f s + 1})S(k_{c1}) \tag{25}$$

where $S(k_{c1})$ represents the sign of controller gain k_{c1} which is included in order to get the correct sign for the controller gain k_{c0}. Also, K_c, τ_I, τ_D and τ_f represent the controller gain, integral time, derivative time and filter time constant respectively, which can be expressed as follows:

$$K_c = \frac{\lambda_1 - 1}{\lambda_1 r_1 \tau_z K_p |K_p| (\tau_p + \tau_z)} \tag{26}$$

where $|K_p|$ denoted the absolute value of K_p.

$$\tau_I = \gamma(1 + r_1 \tau_z) \tag{27}$$

$$\tau_D = \tau_p \tag{28}$$

$$\tau_f = \frac{\tau_p}{\lambda_1} \tag{29}$$

4. Proposed MSC Scheme

The key feature of the proposed scheme is to combine both direct and indirect MSC ideas. Fig. 3 demonstrates the block diagram of the proposed MSC scheme for controlling (i.e., highly nonminimum-phase) integrating unstable process.

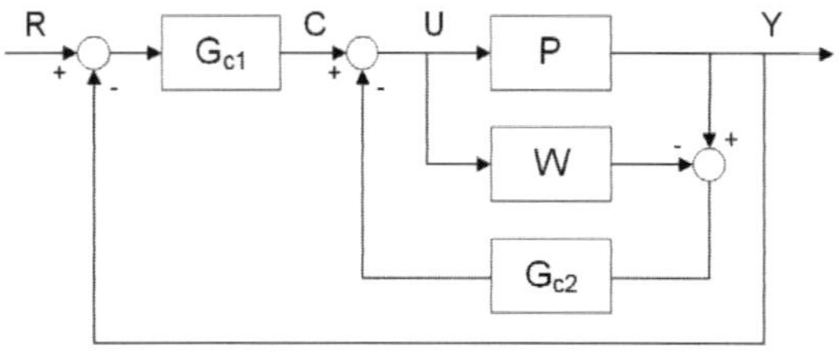

Figure 3. MSC with double-loop scheme

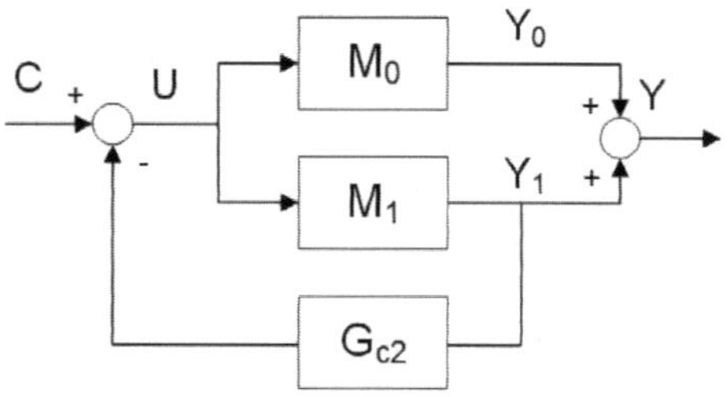

Figure 4. Equivalent structure of the secondary loop

The indirect MSC scheme is used to design the secondary controller G_{c2} in order to first pre-stabilize the unstable process, whilst the direct MSC scheme is used to design the primary controller G_{c1}. Fig. 4 shows the equivalent structure of the secondary loop with the implementation of indirect MSC scheme.

4.1. Tuning Relations for Secondary Controller

Consider a second order unstable integrating process represented in term of

$$P(s) = \frac{K_p(-\tau_z s + 1)}{s(\tau_p s - 1)} \tag{30}$$

where K_p, τ_z and τ_p represent the process gain, lead time constant and time constant respectively. After the decomposition by partial fraction expansion, the sum of two basic modes written as

$$P(s) = \underbrace{\frac{k_{p0}}{s}}_{m_0} + \underbrace{\frac{k_{p1}}{\tau_p s - 1}}_{m_1} \tag{31}$$

where m_0 and m_1 denote the outermost and inner-layer modes, respectively and the mode gains are

$$k_{p0} = -K_p \tag{32}$$

$$k_{p1} = -K_p(\tau_z - \tau_p) \tag{33}$$

Referring to Fig. 4, a transfer function from C to U is given as

$$H_u = \frac{K_u(\tau_p s - 1)}{\tau_{c2} s + 1} \tag{34}$$

A closed-loop transfer function from C to Y_1 can be defined as

$$G_2 = \frac{K_{c1}^0}{\tau_{c2} s + 1} \tag{35}$$

Also, a transfer function from C to Y_0 can be written in the form of

$$G_0 = \frac{K_{c0}^0(\tau_p s - 1)}{s(\tau_{c2} s + 1)} \tag{36}$$

where the parameters in (34)-(36) are given by

$$K_u = \frac{1}{K_{c2} k_{p1} - 1} \tag{37}$$

$$K_{c1}^0 = \frac{k_{p1}}{K_{c2} k_{p1} - 1} \tag{38}$$

$$K_{c0}^0 = \frac{k_{p0}}{K_{c2} k_{p1} - 1} \tag{39}$$

$$\tau_{c2} = \frac{\tau_p}{K_{c2} k_{p1} - 1} = \frac{\tau_p}{-K_{c2} K_p(\tau_z - \tau_p) - 1} \tag{40}$$

The augmented plant transfer function from C to Y is obtained by summing (35) and (36) as follows:

$$P_a(s) = G_0(s) + G_2(s) = H_u(s)P(s) \tag{41}$$

After the simplification of (41), the augmented plant transfer function P_a can be expressed as

$$P_a(s) = \frac{K_{pa}(-\tau_z s + 1)}{s(\tau_{c2} s + 1)} \tag{42}$$

where the augmented process gain is given by

$$K_{pa} = -K_{c0}^0 = \frac{K_p}{-K_{c2}K_p(\tau_z - \tau_p) - 1} \tag{43}$$

It is interesting to note that, the ill-conditioned process of the form given by (30) can now be relieved to (42), i.e. similar to (4), which is relatively easy to stabilize.

4.2. Secondary Controller Setting

Referring to (10), by applying the Routh stability criterion to its characteristic equation to get the range $\tau_{c2} > 0$. Thus, the following limit can be obtained from (40). For simple tuning, let a parameter r_2 be defined as follows:

$$-K_{c2}K_p(\tau_z - \tau_p) = r_2 > 1\,; \quad r_2 > 1 \tag{44}$$

Here, r_2 is used as a tuning parameter to calculate K_{c2}, which ensures local stability of the mode m_1 with a range of $r_2 > 1$ as follows:

$$K_{c2} = \frac{r_2}{-K_p(\tau_z - \tau_p)}\,; \quad r_2 > 1 \tag{45}$$

Also, the augmented process gain in (43) can now be rewritten as

$$K_{pa} = \frac{K_p}{r_2 - 1} \tag{46}$$

A recommended range of $0.5\tau_z \le r_2 \le \tau_z + \tau_p$ is sufficient for the partial stabilization purpose for secondary controller tuning.

4.3. Primary Controller Tuning

The overall system performance is to be attained via the tuning of the main controller G_{c1}, which can be referred to (26)-(29) in Section 3. Meanwhile, the controller parameters based on the partially stabilized plant P_a in (42) can be expressed in the form of

$$K_c = \frac{(\lambda_1 - 1)(r_2 - 1)^3}{\lambda_1 r_1 \tau_z K_p |K_p| (\tau_p + \tau_z)} \tag{47}$$

$$\tau_I = \gamma(1 + r_1\tau_z) \tag{48}$$

$$\tau_D = \tau_p \tag{49}$$

$$\tau_f = \frac{\tau_p}{\lambda_1} \tag{50}$$

From (47), an approximated linear relation which provides an inversely proportional relationship between the controller gain and the process gain is given as follows, i.e., one of the terms $|K_p|$ has been removed.

$$K_c = \frac{(\lambda_1 - 1)(r_2 - 1)^3}{\lambda_1 r_1 \tau_z K_p (\tau_p + \tau_z)} \tag{51}$$

Thus, (48)-(51) represent the MSC-PID tuning relations, which can be tuned by adjusting λ_1, γ, r_1 and r_2.

4.4. Robustness Criteria

Sensitivity function plays an important role for judging the performance-robustness of the system. The maximum peak of sensitivity function in frequency domain is defined as follows:

$$M_s = \|[1 + G_{c1}(j\omega)P_a(j\omega)]^{-1}\|_\infty \tag{52}$$

where the maximum peak of sensitivity function is recommended in the range of $M_s < 2.0$ [18]. The lower value of maximum peak of sensitivity function leads to higher robustness of controller but results in a sluggish response. A range of $1.0 < M_s < 2.0$ is recommended to give a practical response.

4.5. Design Procedure

The general design procedure is generated based on the design of double-loop control structure for unstable integrating systems. The indirect MSC scheme is used to stabilize the unstable process in the secondary loop, while direct MSC scheme is used as the main controller. The design steps are as follows:

Step 1: Tune the secondary controller, i.e. obtain the value of controller gain K_{c2} via (45) by specifying the tuning parameter r_2 with a range of $0.5\tau_z \leq r_2 \leq \tau_z + \tau_p$.

Step 2: Obtain the augmented plant transfer function P_a as (42). The value of K_{pa} and τ_{c2} can be calculated by (40) and (43).

Step 3: The main controller G_{c1} in (25) is tuned by specifying the four MSC tuning parameters, which are λ_1, γ, r_1 and r_2 via (48)-(51) for the controller gain K_c, integral time τ_I, derivative time τ_D and filter time constant τ_f respectively. Note that, $\lambda_1 > 1$, $r_1 > 1$, $0.5\tau_z \leq r_2 \leq \tau_z + \tau_p$ and $0.5 \leq \gamma \leq 3.0$. As a suggestion, set $\lambda_1 = 5$ while altering r_1, r_2 and γ in order to achieve GM close to 7.5 dB - 8.5 dB, PM close to 45°- 60°and maximum peak of sensitivity function in the range of $1.0 < M_s < 2.0$ by referring to (52).

5. Illustrative Example

A real case of fed batch bioethanol production is used as a case study in this paper [19]. A linearized open-loop second order unstable integrating process (SOUIP) is used to demonstrate the effectiveness of the proposed control scheme as follows:

$$P(s) = \frac{-13.99(-3.81s + 1)}{s(3.83s - 1)} \tag{53}$$

Based on (30), the process parameters are stated as: $K_p = -13.99$, $\tau_z = 3.81$ and $\tau_p = 3.83$. The $-\tau_z$ and negative pole value causes the inverse response and unstable open-loop system respectively. Note that, the process is extremely difficult to stabilize with the conventional PID controller including with some of the advanced control techniques, e.g., Linear-Quadratic Gaussian (LQG) and robust control. So far, there is no report in the open literature of a control scheme which can stabilize such a type processes.

The design procedure of the proposed control system design is shown in Section 4.5. The secondary controller is obtained by specifying the tuning parameter $r_2 = 7.5$ in order to pre-stabilize the unstable process, which leads to $K_{c2} = -29.14$ by using (45). For the main controller, the finalized MSC tuning values are $\lambda_1 = 5$, $r_1 = 3$ and $\gamma = 2$, which leads to GM = 8.04dB and PM = 46.1°. Thus, the MSC-PID controller is given by

$$G_c(s) = -0.0481\left(1 + \frac{1}{24.8480s}\right)\left(\frac{0.5887s + 1}{0.1177s + 1}\right) \tag{54}$$

The maximum peak of sensitivity function is given as $M_s = 1.67$, which is within the recommended range. Note that, a set point pre-filter (F_r) is suggested in order to reduce the overshoot response in setpoint tracking. The setpoint pre-filter is expressed by

$$F_r = \frac{\frac{\tau_I}{3}s + 1}{\tau_I s + 1} \tag{55}$$

In order to compare the performances of proposed MSC scheme, a Skogestad IMC (SIMC) tuning are designed based on a double-loop control scheme [20]. The PID with filter based on double-loop SIMC tuning is shown as follows:

$$G_c(s) = -0.00034\left(\frac{2.2s + 1}{s}\right)\left(\frac{35s + 1}{1.1s + 1}\right) \tag{56}$$

The performances of the proposed control system design are presented based on the response of setpoint tracking, output disturbance and input distrubance with 1 unit step changes. In Fig. 5 the settling time of the proposed MSC scheme is much improved with the employ of setpoint pre-filter and SIMC tuning, i.e. 63 units, which gives a smooth response and faster settling time. Noted that, the setpoint pre-filter is able to reduce the overshoot and underdamped responses in setpoint tracking, which does not affect the disturbances rejection performance. Fig. 6 and Fig. 7 demonstrates the output and input disturbance rejection responses respectively. Obviously, the proposed MSC scheme shows the improved performance compared to the established control strategy, i.e. SIMC tuning.

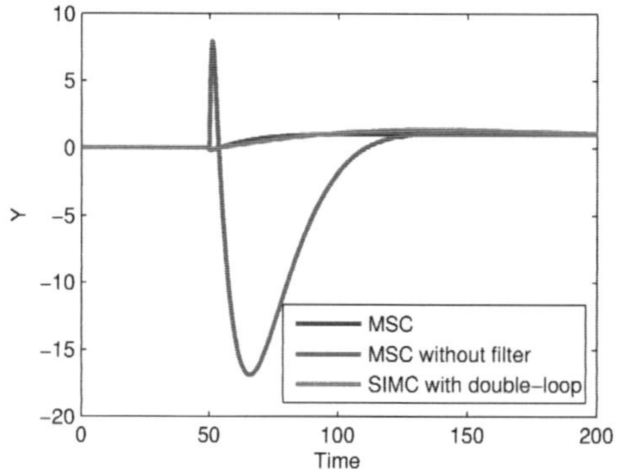

Figure 5. Nominal response for setpoint tracking

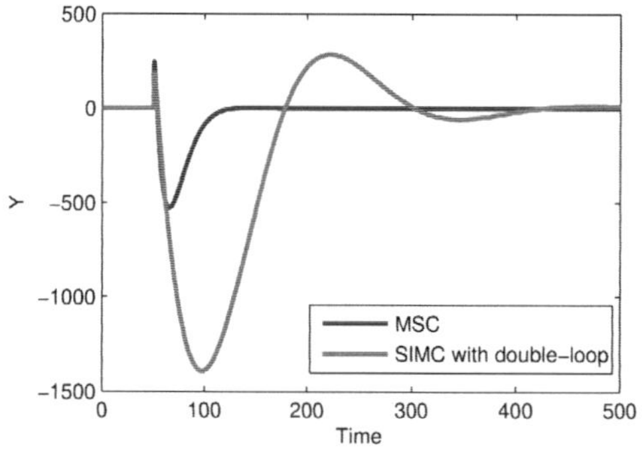

Figure 6. Nominal response for output disturbance rejection

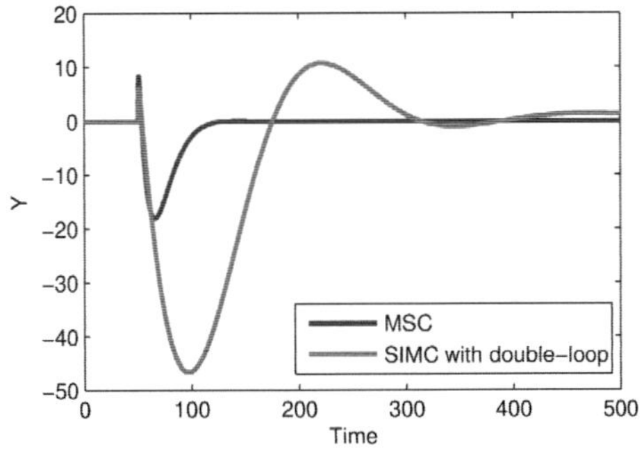

Figure 7. Nominal response for input disturbance rejection

6. Conclusions

In this paper, a new variant of MSC scheme with double-loop control structure has been presented for nonminimum-phase systems with both unstable and integrating modes. The NMP integrating-unstable system is difficult to control even by using some existing control scheme, i.e. well-known LQG controller. However, the proposed control scheme provides an effective scheme using two controllers: a PID controller in the external loop and a P controller in the inner loop. Easy to follow procedure is also given for simple tuning of these two controllers. It is interesting to note that, the performance of setpoint tracking is relatively enhanced with the implementation of setpoint pre-filter for proposed scheme. The directions of the future works will be extended to higher-order unstable and integrating processes.

Acknowledgments

This work is supported by a grant from the Curtin Sarawak Research Institute (CSRI).

References

[1] Park J H, Sung S W and Lee I B 1998 *Automatica* **34** 751–756
[2] Panda R C 2009 *Chemical Engineering Science* **64** 2807–2816
[3] Wang Y G and Cai W J 2002 *Industrial & engineering chemistry research* **41** 2910–2914
[4] Rotstein G E and Lewin D R 1992 *Computers & chemical engineering* **16** 27–49
[5] Alford J S 2006 *Computers and Chemical Engineering* **30** 1464–1475
[6] Rotstein G E and Lewin D R 1991 *Industrial & engineering chemistry research* **30** 1864–1869
[7] Lee Y, Lee J and Park S 2000 *Chemical Engineering Science* **55** 3481–3493
[8] Poulin E and Pomerleau A 1996 *Control Theory and Applications* **143** 429–435
[9] Paor A M D 1985 *International Journal of Control* **41** 1025–1036
[10] Paor A M D and O'Malley M 1989 *International Journal of Control* **49** 1273–1284
[11] Chen Y and Won S 2008 *International Conference on Control, Automation and Systems 2008* pp 618–622
[12] Simhachalam D and Mudi R K 2014 *2014 International Conference on Control, Instrumentation, Energy & Communication(CIEC)* pp 51–55
[13] Nandong J and Zang Z 2014 *Journal of Process Control* **24** 600 – 612
[14] Nandong J and Zang Z 2013 *Journal of Process Control* **23** 1333–1343
[15] Nandong J and Zang Z 2013 *Industrial & Engineering Chemistry Research* **52** 8248–8259
[16] Nandong J and Zang Z 2013 *Industrial Electronics and Applications (ICIEA), 2013 8th IEEE Conference on IEEE* pp 1527–1532
[17] Ugon B, Nandong J and Zang Z 2014 *Industrial Electronics and Applications (ICIEA), 2014 IEEE 9th Conference* pp 775–780
[18] Panagopoulos H and Aström K J 2000 *International Journal of Robust and Nonlinear Control* **10** 1249–1261
[19] Sonnleitnert B and Käppeli O 1986 *Biotechnology and bioengineering* **28** 927–937
[20] Skogestad S 2003 *Journal of Process Control* **13** 291 – 309

Simulation and control of water-gas shift packed bed reactor with inter-stage cooling

S Z Saw[1,2] and J Nandong[2,3]

[1] Curtin Sarawak Research Institute, Curtin University Sarawak, Miri 98009, Malaysia
[2] Department of Chemical Engineering, Curtin University Sarawak, Miri 98009, Malaysia

E-mail: shuey.zi@postgrad.curtin.edu.my,jobrun.n@curtin.edu.my

Abstract. Water-Gas Shift Reaction (WGSR) has become one of the well-known pathways for H_2 production in industries. The issue with WGSR is that it is kinetically favored at high temperatures but thermodynamically favored at low temperatures, thus requiring careful consideration in the control design in order to ensure that the temperature used does not deactivate the catalyst. This paper studies the effect of a reactor arrangement with an inter-stage cooling implemented in the packed bed reactor to look at its effect on outlet temperature. A mathematical model is developed based on one-dimensional heat and mass transfers which incorporate the intra-particle effects. It is shown that the placement of the inter-stage cooling and the outlet temperature exiting the inter-stage cooling have strong influence on the reaction conversion. Several control strategies are explored for the process. It is shown that a feedback-feedforward control strategy using Multi-scale Control (MSC) is effective to regulate the reactor temperature profile which is critical to maintaining the catalysts activity.

1. Introduction

The potential of Hydrogen is being widely examined as a possible energy carrier because of its high energy content and environmental compatibility. Hydrogen can be produced from a wide range of primary energy sources and pathways. Presently, Water-Gas Shift Reaction (WGSR) represents an attractive pathway for H2 production and this reaction is currently used in many important industrial applications [1].

The WGSR is a well-known and extensively studied reaction, which is a mildly exothermic and equilibrium limited reaction [2] shown in equation (1):

$$CO + H_2O \rightleftharpoons H_2 + CO_2 \quad (1)$$

The equilibrium of the WGSR favours low temperatures and is unaffected by the operating pressure. In the first stage of the industrial-scale process units, a high temperature (HT-WGS) packed bed reactor is used to exploit fast kinetic rates. The outlet stream of the HT-WGS reactor is cooled and fed to low temperature (LT-WGS) packed bed reactor where the temperature is beneficial towards the equilibrium henceforth leading to a high CO conversion [3] (see figure 1).

The goal of the present study is to find an answer to a question: is it possible to achieve higher CO conversion with only one stage (HT-WGS) packed bed reactor and moreover, to

[3] Corresonding author. Tel.:+60-85-44-3939

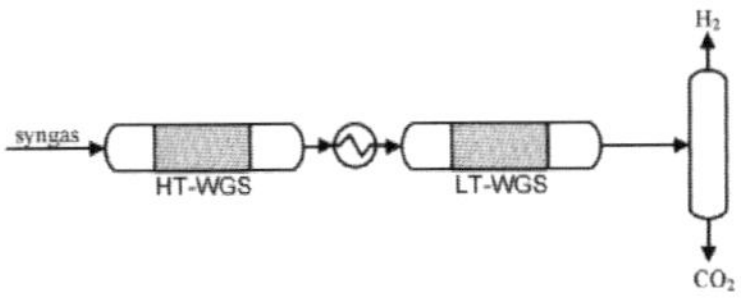

Figure 1. Conventional process configuration for high and low temperature packed-bed WGS reactors including separation of CO_2 and H_2.

design a simple control strategy to control the reactor to maintain its optimum performance. Furthermore, since the WGSR is exothermic in nature, the reaction is often conducted in adiabatic shift reactors to avoid catalyst overheating as well as to improve the reaction conversion [4]. Then again, an excessively high temperature can cause the catalysts activity to degenerate and in turn leads to a poor conversion of CO. For this reason, an effective temperature control of the reactor in which the WGSR takes place is very important to avoid serious catalyst deactivation.

The WGSR has been extensively studied in several works, both experimentally and using numerical simulations. Simulation is a very useful tool for predicting and studying these reactors, but the results obtained are highly dependent on the accuracy of the models used [5]. Most models proposed in the literature, (e.g. [6], [7] and [8]), often disregard the influence of intra-particle mass and heat transfer effects, although they can be important because of fast kinetics of this reaction.

With respect to control strategies in a packed bed tubular reactor (PBTR) where complex reactions such as the WGSR takes place, a single-loop controller is hardly effective for controlling the reactors effluent temperature due to significant transport delay along the reactor. In this case, a feed-forward control strategy could be employed to reject the disturbance before it can greatly upset the performance of the reactor. Even so, feedforward controller only rejects a specific measured disturbance and not all of the disturbances that occur in a process. For this reason, it is a common in industry that the feedforward controller is used together with feedback control system against certain disturbances [9]. Interestingly, both the feedback and feedforward controllers can be effectively designed based on the recently proposed multi-scale control (MSC) scheme in [10].

In this study, we focus on one stage (HT-WGS) packed bed reactor with inter-stage cooling. A rigorous dynamic distributed model is developed based on the first principles that integrate both the intra-particle heat and mass transfers. In addition, three control strategies (i.e., single loop control, conventional feedforward and feedback control, and MSC feedforward and feedback control schemes) will be implemented to regulate the reactor temperature at a selected location of the WGS reactor to achieve: (1) high CO conversion and (2) to suppress the temperature from reaching the threshold value of catalyst deactivation temperature.

The paper is structured as follows: the next section (section 2) includes the WGSR kinetics and HT-WGS packed bed tubular reactor (PBTR) modelling. Section 3 presents the formulation of control strategies. Section 4 provides the simulation results and discussion. Finally, Section 5 highlights some conclusions and future research directions.

2. WGSR modelling

2.1. WGSR kinetics

The detailed explanation on kinetics and reactor modelling of WGSR can be found in [1]. Here, we only use Power Law (PL) model to describe the kinetics as it is the simplest empirical rate expression, which disdain any mechanism such as redox or Langmuir-Hinshelwood (LH). Likewise, several researchers have been reported using this kinetic model for high temperature

WGSR, e.g., [11] and [12]. The PL model is given by:

$$-r_A = k_0 e^{-E_a/RT} N_{CO}^a N_{H_2O}^b N_{CO_2}^c N_{H_2}^d (1-\beta) \tag{2}$$

where, r_A (mol/g.hr) is the rate of reaction, N_i (mol/hr) is the corresponding molar flow rate of component i whereas, a, b, c, and d are the apparent reaction orders of component CO, H_2O, CO_2, and H_2, respectively; k_0 is the pre-exponential factor, E_a is the activation energy (J/mol), R is the universal gas constant (8.314 J/mol.K), and T is the reaction temperature (K). is the term for backward reaction and is defined as follows:

$$\beta = \frac{N_{CO_2} N_{H_2}}{K_P N_{CO} N_{H_2O}} \tag{3}$$

where, K_P is an equilibrium constant for WGS reaction and the equation for K_P calculation is as below:

$$\ln K_P = \left(\frac{4577.8}{T}\right) - 4.33 \tag{4}$$

The dimension of the PBTR system used in this study is scaled to 1 (one) metre in length with additional 0.1m for inter-stage cooling and 16cm in diameter (D_t) based on the pilot-scale membrane reactor commercialized by [13]. For the PBTR modelling, Fe_3O_4-Cr_2O_3 catalyst is used with its property as reported in [5]. For the purpose of dynamic modelling, the length of the reactor is sub-divided into eleven equal parts. One of the parts is the inter-stage cooling as shown in figure 2 where L_1 and L_3 are subdivided into 5 equal lengths of 0.1m. L_2 represents the inter-stage cooling in-between the reactor. Figure 3 shows the cross section of the inter-stage cooling where the tube is arranged in triangular (30°) pattern as this pattern manages to produce high turbulence and consequently results in high heat transfer coefficient [14].

2.2. Packed bed tubular reactor model with inter-stage cooling

In order to estimate the outlet temperature of the reactor, a theoretical numerical model is used for modelling the PBTR. This model consists of mass and energy balance equations with the incorporation of intra-particle mass and heat transfers. The assumptions that govern these equations are: (1) only the axial temperature change is taken into account, (2) the radial temperature change is neglected due to the fact that the diameter of the reactor is small and adiabatic (no heat loss from the surface of the reactor); see [15], (3) momentum conservation equation has been disregarded in this model because the bed height is too small to create a significance amount of pressure drop, i.e., see [16], (4) no reaction in the inter-stage cooling, and (5) inter-stage cooling consists of several small tubes therefore radial temperature change

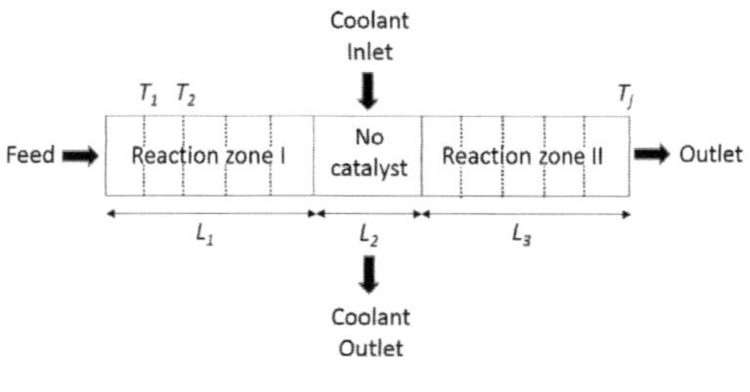

Figure 2. Division of the packed bed tubular reactor into 11 sub-sections.

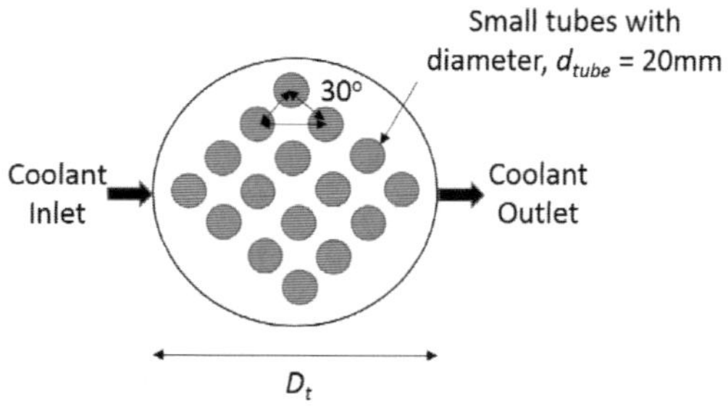

Figure 3. Cross-sectional diagram of the inter-stage cooling L_2 in PBTR.

is neglected. Equations (5) - (7) show the mass and energy balance equations for the PBTR system.

$$\epsilon V \frac{dC_{i,j}}{dt} = F(C_{i,j-1} - C_{i,j}) \pm r_{A,j}\eta_{gl}W \tag{5}$$

$$(1-\epsilon)V\rho_s C_{p,s}\frac{dT_{s,j}}{dt} = r_{A,j}\eta_{gl}W\Delta H_r - UA(T_{s,j} - T_{f,j}) \tag{6}$$

$$\epsilon V\rho_f C_{p,f}\frac{dT_{f,j}}{dt} = FC_{p,f}\rho_f(T_{f,j-1} - T_{f,j}) + UA(T_{s,j} - T_{f,j}) \tag{7}$$

Here, ϵ denotes the bed voidage factor, V (m^3) the sub-section volume of the reactor, $C_{i,j}$ (mol/m^3) the molar concentration of component i at the j-th sub-section, F (m^3/hr) the feed flow rate, W (g) the weight of catalyst in the reactor sub-section, ρ_s and ρ_f (kg/m^3) the density of catalyst and fluid respectively, $C_{p,s}$ and $C_{p,f}$ (kJ/kg.K) the specific heat capacity of catalyst and fluid, $T_{s,j}$ (K) the catalyst temperature at j-th sub-section, $T_{f,j}$ (K) the fluid temperature at j-th sub-section, ΔH_r (kJ/mol) the heat of reaction, U (kJ/hr.m^2.K) the convective overall heat transfer coefficient, A (m^2) the heat transfer area and η_{gl} is the global effectiveness factor.

In equation (5), the $\pm$ sign indicates either species consumption or generation; for CO and H_2O the sign is minus, while for the products the sign is plus. As for the inter-stage cooling, the temperature profile of the coolant and the feed will be calculated based on energy balance model as follows:

$$\epsilon V\rho_f C_{p,f}\frac{dT_{f,j}}{dt} = FC_{p,f}\rho_f(T_{f,j-1} - T_{f,j}) - hA_t(T_{f,j} - T_{c,out}) \tag{8}$$

$$V_c\rho_c C_{p,c}\frac{dT_{c,out}}{dt} = F_c C_{p,c}\rho_c(T_{c,in} - T_{c,out}) + hA_t(T_{f,j} - T_{c,out}) - F_c\rho_c L_c \tag{9}$$

where all the subscript c stands for coolant, h (kJ/hr.m^2.K) is the overall heat transfer coefficient, At (m^2) is the area of heat transfer in the inter-stage cooling and L_c (kJ/kg) is the latent heat of vaporization. The h is calculated based on Gnielinski correlation [17] and Sieder and Tate correlation [18].

The global effectiveness factor takes into account both mass transfer resistances, internal and external to the catalyst particles. In the global effectiveness calculation, the effective diffusion coefficient used is approximated by Knudsen diffusion in accordance with the physical properties of the gas compounds and textural properties of the catalyst. Meanwhile, the mass transfer resistance external to the catalyst particle is modelled using the mass transfer coefficient for transport in packed bed and the mass transfer resistance internal to the catalyst particle is modelled using generalized Thiele modulus. The internal effectiveness factor is used to model the contribution of the mass transfer resistance inside the porous catalyst particles. The expression of the internal effectiveness factor is suitable for a catalyst in the form of pellets and expression of the used in this equation can deal with the reversible power-law type kinetic equation. The equations mentioned above are written as follows.

$$\eta_{int} = \frac{1}{\Phi}\left(\frac{1}{tanh(3\Phi)} - \frac{1}{3\Phi}\right) \tag{10}$$

$$\Phi = \frac{d_p/4\rho_s r_{A,j}}{\sqrt{2D_{ep}\rho_s\int_{C_{eq}}^{C_i} r_{A,j}dC_i}} \tag{11}$$

$$D_{ep} = \frac{d_{pore}\epsilon_{int}}{\tau}\frac{4}{3}\sqrt{\frac{RT}{2\Pi M}} \tag{12}$$

Table 1. Fixed numerical parameters.

Name	Numerical Parameters	
	Value	Reference
Length of reactor, L	1 m	[13]
Diameter of reactor. D_t	16 cm	[13]
Catalyst density, ρ_s	1945 kg/m^3	[5]
Catalyst diameter, d_p	2×10^{-3} m	[5]
Catalyst pore diameter, d_{pore}	9×10^{-9} m	[5]
Catalyst bed porosity, ϵ	0.4	[5]
Catalyst tortuosity, τ	4	[5]
Heat of reaction, ΔH_r	-41.1 kJ/mol	[3]
Feed flowrate	16 m^3/hr	[3]
Steam/Carbon (S/C) ratio	2.4	[3]
Inlet temperature, T_{in}	873 K	[3]
Coolant inlet temperature, $T_{c,in}$	300 K	[3]

$$a_s K_G = \frac{6D_{COm}}{d_p^2}(2 + 1.1Re^{0.60}Sc^{0.33}) \tag{13}$$

$$\eta_{gl} = \frac{1}{\frac{1}{\eta_{gl}} + \frac{\rho_s}{a_s K_G}\left(\frac{\partial r_{A,j}}{\partial C_{CO}}\right)} \tag{14}$$

where Φ denotes the generalized Thiele modulus, d_p (m) is the diameter of the catalyst particle, D_{ep} is the pore effective diffusion coefficient, $a_s K_G$ denotes the mass transfer coefficient, d_{pore} (m) is the diameter particle pore size, ϵ_{int} is the internal porosity, is the tortuosity, M (kg/mol) is molar weight, D_{COm} (m^2/s) is molecular diffusion of CO in the reaction mixture, Re is Reynolds number and Sc is Schmidt number.

The PBTR model is solved using MATLAB (v. 2013b) software. Not to mention that the software has been used to model packed bed reactors with satisfactory results in previous works [19]. Notice that in this work we sub-divide the reactor length into 11 sub-sections of equal lengths to capture the effects of temperature and concentration distributions along the reactor (the actual model should be in PDE form). Hence, a set of ordinary differential equations (ODEs) is used instead of partial differential equations (PDEs). All the numerical parameters used in this simulation are listed in Table 1.

3. Control strategies

3.1. Single loop feedback control strategy

Single loop control strategy consists of a primary process P_1 and a controller G_{c1} in a single closed loop. The controller for this single loop strategy is tuned based on the model obtained by linearizing the nonlinear PBTR described in equations (5) - (7). The transfer function is obtained through linearizing the control variable against the manipulated variable which in this case is the flowrate of the inter-stage coolant. Equation (15) shows the transfer function obtained from linearizing the abovementioned variables.

$$H(s) = \frac{Y_1(s)}{R(s)} = \frac{G_{c1}P_1}{1 + G_{c1}P_1} \tag{15}$$

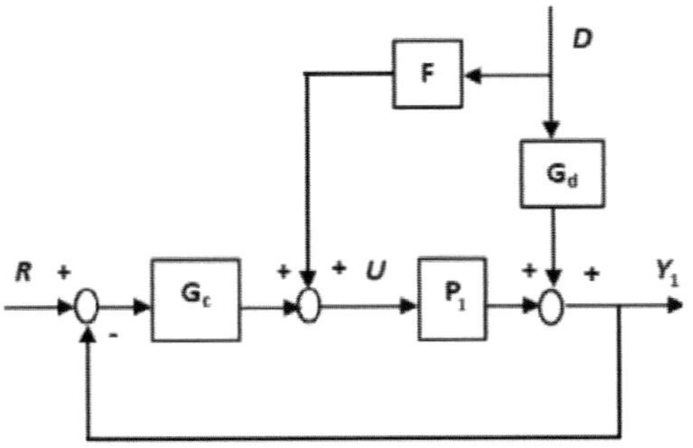

Figure 4. Conventional feedback-feedforward control structure.

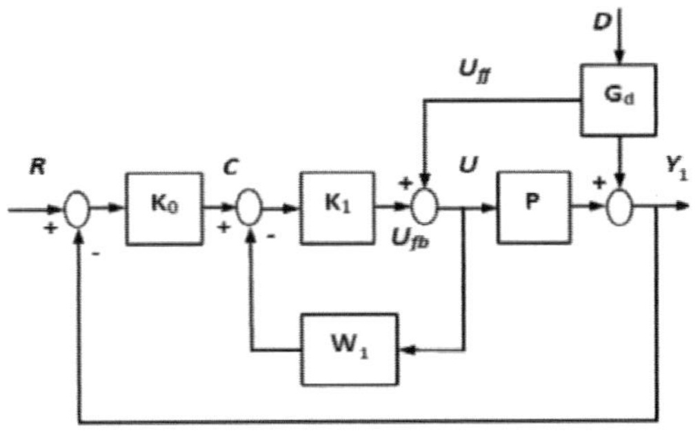

Figure 5. Multi-scale feedback-feedforward control structure.

3.2. Generalized feedback-feedforward control strategy

Figure 4 depicts the block diagram of a conventional feedback plus feedforward control structure commonly used in process industry. Here, G_c is the feedback controller; F the feedforward controller.

In figure 4, the feedforward controller F is derived by inverting the nominal model P_1 with an assumption that, the output disturbance effect via G_d is cancelled out by the feedforward controller (ideal design), thus:

$$[F(s)P_1(s) + G_d(s)]D(s) = 0 \tag{16}$$

So, the ideal feedforward controller F is obtained as:

$$F(s) = -\frac{G_d(s)}{P_1(s)} \tag{17}$$

3.3. MSC feedforward-feedback control strategy

Figure 5 illustrate the block diagram of the proposed feedback-feedforward control structure based on the multi-scale control (MSC) scheme [9]. The feature that distinguishes this MSC-based feedback-feedforward (MSC FB-FF) control structure from the conventional one (figure 4) is that the MSC-FB/FF treats the measured disturbance D as though it is an input disturbance introduced directly into the fast inner-loop of the MSC control system. In this way, the effect of disturbance D on Y is mainly removed by the fast inner-loop sub-controller K_1.

The details of the MSC-FB/FF control strategy can be obtained from [9]. In figure 4, the transfer function from U_{ff} to U is given by

$$H_{fu}(s) = \frac{1}{1 + K_1 W_1(s)} \tag{18}$$

where K_1 is the inner-layer sub-controller which is often chosen as a P-only controller because this leads to simple controller tuning and W_1 is a multi-scale predictor. Meanwhile, a PI controller is often chosen for the outermost sub-controller K_0; an integral action is added to remove steady-state offset in the output.

The MSC FB-FF structure (figure 5) can be reduced to the same structure as that of the conventional single-loop feedback-feedforward control. One can derive an equivalent MSC-based feedforward controller as follows:

$$F_{msc}(s) = H_{fu}(s)F^*_{ss}(s) \tag{19}$$

Equation (19), F^*_{ss} denotes the optimised static feedforward controller while F_{msc} denotes the MSC-based feedforward controller, i.e., the subscript msc is used to differentiate it from the conventional feedforward controller in figure 4. The formula for obtaining static the feedforward is obtained via the proposed method in [9].

4. Results and discussion

4.1. Numerical parameters and model verification

The fixed numerical parameters used in this study are listed in Table 1. In Table 1, catalyst bed properties and reaction kinetic parameters are taken from the study of [5]. The operating conditions are based on the study of [3]. As shown in [20] which is the extended study from [5], the numerical model was verified with the experimental data in [20]. Using the data listed in Table 1, the numerical model is verified again to ensure its validation.

Meanwhile, figure 6 shows the comparison between the numerically predicted CO conversions for the packed-bed reactor with experimental data reported in [1]. As shown in figure 6, the agreement between numerical model and experimental data is good.

4.2. Effects of inter-stage cooling

Based on the verification discussed above, the numerical model is adequately extended to study the PBTR under inter-stage cooling shown in figure 2. Water coolant is used in this work. Using the fixed parameters and base operating conditions listed in Table 1, figure 7 show the reactor conversion comparison between traditional PBTR and PBTR with inter-stage cooling. For traditional PBTR, a maximum conversion is reached at the 5th reactor section and an increase in the length of the reactor does not improve the conversion as WGS is a reversible reaction equilibrium limited.

On the contrary, if the inter-stage cooling is placed right after the maximum conversion is achieved, it can be seen from figure 7 a new maximum conversion can be attained. This inter-stage cooling will not only bring out a better conversion, it will also reduce the risk of catalyst deactivation. The temperature decreases to a lower temperature just before the reactor temperature reaches the catalyst deactivation temperature, which is T_{deac} =1200K for the catalysts used in this work.

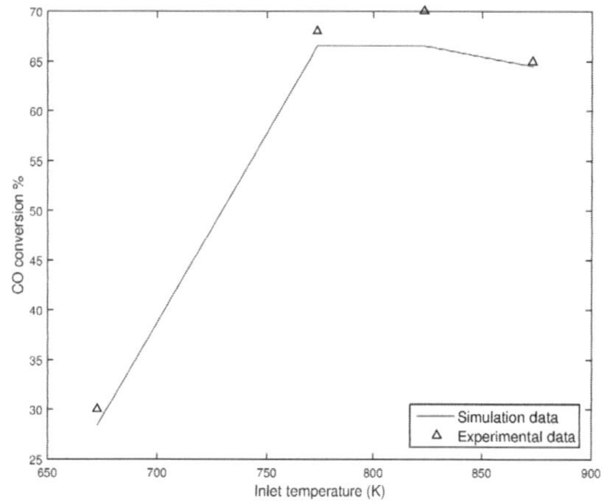

Figure 6. Numerical model verification by comparing computed CO conversion with experimental data reported by [21].

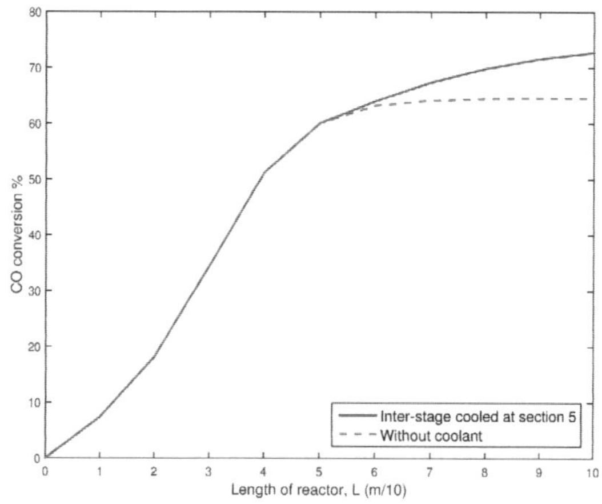

Figure 7. Comparison of CO conversion between traditional PBTR and PBTR with inter-stage cooling.

4.3. Effects of inter-stage cooling placement

Further investigation on the effect of placement of the inter-stage cooling on the PBTR performance is also examined. As shown in figure 8, these different sections referred to the placement of L_2.

To illustrate, with the inter-stage cooling at section 3, $L_1 = 0.3$m followed by $L_2 = 0.1$m and $L_3 = 0.7$m. Note that, the combination of L_1, L_2 and L_3 should only give a value of 1.1m in total reactor length.

Based on the results shown in figure 8, the placement of inter-stage cooling influences the CO conversion. The maximum CO conversion is obtained when the inter-stage cooling is placed around sections 5 and 6. Implementation of inter-stage cooling at other than 5th and 6th sections were observed to accomplish lower CO conversion but still higher compared to the traditional PBTR. This is because inter-stage cooling placed at section 5 and 6 has already achieved a maximum conversion at the first part of the reactor, L_1 with ample of length in L_3 reactor for the reaction to occur again.

4.4. Inter-stage cooling outlet temperature effect

The outlet temperature mentioned in this section would be the temperature exiting L_2 just before entering reactor L_3. Figure 9 shows the impact of outlet temperature of L_2 on reactor conversion. The outlet temperature of L_2 is set to be manipulated around these three temperatures (i.e. the initial temperature, 1st section, and 2nd section reactor temperature). Figure 9 show that by dropping the outlet temperature of L_2 equivalent to the 1st section reactor temperature, it will result in the highest CO conversion. The fastest reaction kinetics can be achieved at this particular temperature.

This shows that the outlet temperature of L_2 which also corresponds to the entrance temperature of reactor zone L_3 is very important in terms of thermal management. This behaviour of the reactor motivates the need for a control strategy that is able to control and maintain the outlet temperature of L_2 in order to achieve maximum conversion. Not to mention, the control strategy must be able to suppress temperature below the maximum catalyst temperature in the presence of disturbances.

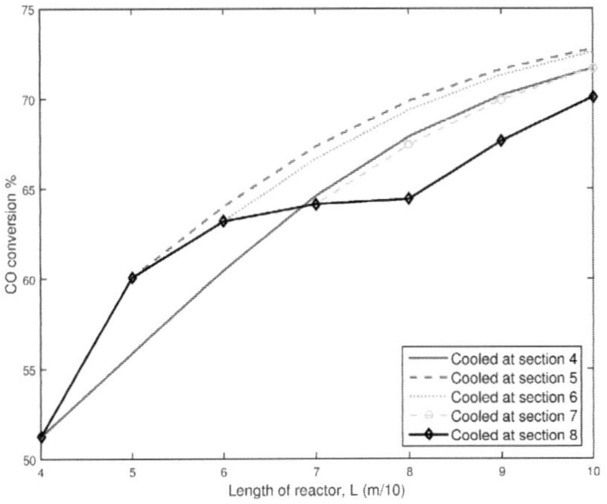

Figure 8. Comparison of CO conversion between different placements of inter-stage cooling at length $L_3 = 0.4$m to 1m.

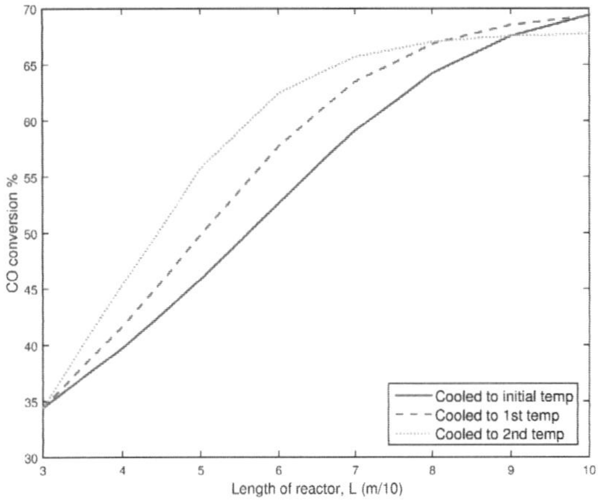

Figure 9. Comparison of CO conversion between different outlet L_2 temperatures for inter-stage cooling at section 3.

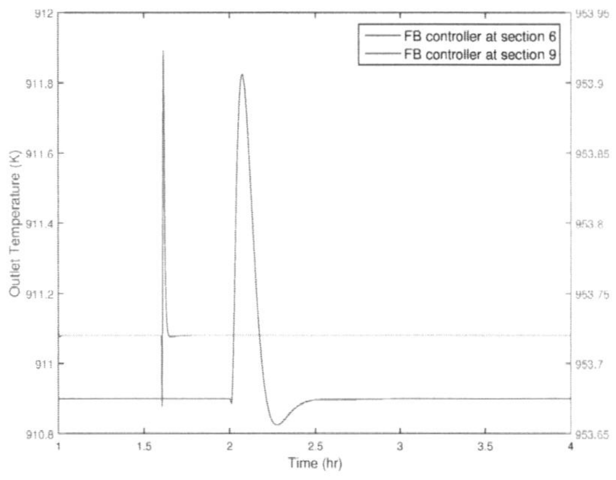

Figure 10. Comparison of overshoot for two different placement of a feedback control strategy.

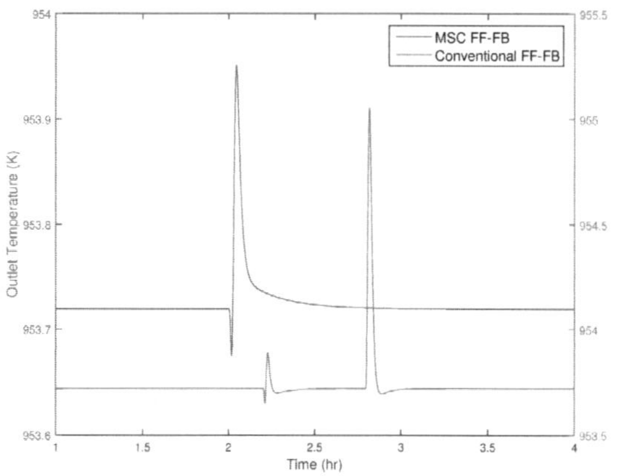

Figure 11. Comparison of MSC FF-FB control with conventional FF-FB control scheme.

4.5. Placement of feedback controller

The effect of the placement of feedback controller is explored under the influences of the inlet feed temperature of 10°C (input disturbance source) while the controlling variable is the flow rate of the inter-stage coolant.

It is shown in figure 11 that the placement of the feedback control at different sections of the reactor plays a significant role in disturbance rejection. In this case, the best and the worst placement of the controller is compared and the conclusion drawn is that the controller at sub-section 9 shows 5% of overshoot, while controller at sub-section 6 shows 15% of overshoot.

4.6. Feedback control vs. generalized FF-FB control vs. MSC FF-FB control strategy

The performance of a single-loop feedback control scheme is compared with both the conventional FF-FB and MSC FF-FB control strategy scheme under the same type of disturbance with additional disturbance of ratio concentration of CO/H_2O.

The feedback controller (primary measurement) is located at sub-section 9 because the placement here shows the best control performance in single-loop control which has been discussed in Section-D. However, the feedforward controller in this work only targeted inlet feed temperature disturbance. In this case study, it is found that conventional feedforward-feedback control scheme show little improvement from feedback control strategy. Therefore, to further improve the control scheme in order to maintain the optimum temperature, multi-scale control (MSC) strategy is implemented into the conventional FF-FB control scheme.

Figure 13 shows the nominal closed-loop response of the outlet fluid temperature at section 9 when the system is subjected to sequential step changes: (1) magnitude of 10 in feed temperature at t = 2 and (2) magnitude of 3 in CO concentration at t = 3. It can be seen that MSC FF/FB has produced significant improvement over the conventional FF/FB control strategy. The overshoot has been drastically reduced by 350%.

5. Conclusion

By introducing an inter-stage cooling in a high-temperature PBTR for WGSR, it can significantly improve the CO conversion. In this paper, it is shown that there is an optimum placement of the inter-stage cooling which is around the middle section of the PBTR. An interesting finding is that the effect of the temperature exiting the inter-stage on the CO conversion is significant.

It is shown that the exit temperature has to be roughly the same as the 1st section of the reactor for an increased maximum conversion to be achieved. Thus, care has to be taken to ensure that the inter-stage outlet temperature does not fluctuate much in the presence of disturbance. Hence, this has suggested that there is a need to find a good control strategy for this process system. Further, we have found out that there is an optimum placement of the conventional feedback controller (primary measurement) located roughly at the upper middle-end of the reactor, at sub-section 9. By using the optimum placement of the feedback controller, a feedforward controller has been combined together with the feedback to improve the reactor temperature control performance. However, the conventional FF-FB control strategy is still lacking in terms of controlling the overshoot. Interestingly, this has been improved via the implementation of the MSC-based FF-FB control strategy. In future works, we will extend the current study to study and develop more complex control strategies for different configurations of the WGSR systems.

Acknowledgement
This work is supported by the Fundamental Research Grant Scheme (FRGS) of the Malaysian Ministry of Higher Education (no: JPT.S Jld.13(28)) and Curtin Sarawak Research Institute grant (no: CSRI 6002).

References
[1] Kumar P, Akpan E, Ibrahim H, Aboudheir A and Idem R 2008 *Industrial & Engineering Chemistry Research* **47** 4086–4097
[2] Adams T A and Barton P I 2009 *International Journal of Hydrogen Energy* **34** 8877–8891
[3] Georgis D, Lima F V, Almansoori A and Daoutidis P 2014 *Industrial & Engineering Chemistry Research* **53** 7461–7469
[4] Mendes D, Mendes A, Madeira L, Iulianelli A, Sousa J and Basile A 2010 *Asia-Pacific Journal of Chemical Engineering* **5** 111–137
[5] Marín P, Díez F V and Ordóñez S 2012 *International journal of hydrogen energy* **37** 4997–5010
[6] Adrover M E, López E, Borio D O and Pedernera M N 2009 *Chemical Engineering Journal* **154** 196–202
[7] Markatos N, Vogiatzis E, Koukou M and Papayannakos N 2005 *Chemical Engineering Research and Design* **83** 1171–1178
[8] Rui Z, Zhang K, Li Y and Lin Y 2008 *International Journal of Hydrogen Energy* **33** 2246–2253
[9] Nandong J 2015 *International Journal of Control, Automation and Systems* **13** 91–98
[10] Nandong J and Zang Z 2013 *Journal of Process Control* **23** 1333–1343
[11] Keiski R L, Salmi T, Niemistö P, Ainassaari J and Pohjola V J 1996 *Applied Catalysis A: General* **137** 349–370
[12] Koryabkina N, Phatak A, Ruettinger W, Farrauto R and Ribeiro F 2003 *Journal of Catalysis* **217** 233–239
[13] Mori T 2005 *Proceedings of the international hydrogen energy congress and exhibition IHEC*
[14] Mukherjee R *et al.* 1998 *Chemical Engineering Progress* **94** 21–37
[15] Aboudheir A, Akande A, Idem R and Dalai A 2006 *International journal of hydrogen energy* **31** 752–761
[16] Shahrokhi M and Baghmisheh G 2005 *Chemical Engineering Science* **60** 4275–4286
[17] Namburu P K, Das D K, Tanguturi K M and Vajjha R S 2009 *International Journal of Thermal Sciences* **48** 290–302
[18] Whitaker S 1972 *AIChE Journal* 361–371
[19] Marin P, Hamel C, Ordonez S, Diez F, Tsotsas E and Seidel-Morgenstern A 2010 *Chemical Engineering Science* **65** 3538–3548
[20] Sanz R, Calles J, Alique D, Furones L, Ordóñez S and Marín P 2015 *International Journal of Hydrogen Energy* **40** 3472–3484

IOP Conf. Series: Materials Science and Engineering **121** (2016) 012023 doi:10.1088/1757-899X/121/1/012023

CO_2 storage in heterogeneous aquifer: A study on the effect of injection rate and $CaCO_3$ concentration

A Raza[1*], R Rezaee[2], C H Bing[1], R Gholami[1], R Nagarajan[1] and M A Hamid[1]
[1] Faculty of Engineering and Science, Curtin University, Miri, Sarawak, Malaysia
[2] Faculty of Science and Engineering, Curtin University, Bentley WA 6102, Australia

*Email: arshadraza212@gmail.com

Abstract. Trapping mechanisms taken place during and after CO_2 injection in a geologic storage medium are impacted by a number of parameters including injection rates together with rock and pore fluid compositions. There have been many studies on the factors controlling the capillary trapping and injectivity of CO_2 storage sites. However, there are only few works carried out discussing on the effect of flow rate and rock and fluids compositions in controlling the trapping mechanisms. In this paper a CO_2 storage site located in a heterogeneous aquifer is simulated to investigate the efficiency of structural, capillary and dissolution trappings as a function of injection rate and concentration of calcium carbonate. The results obtained from numerical analysis indicated that CO_2 injection must be made within an optimum injection rate for having an effective storage in place. It was also found that concentration of $CaCO_3$ is an important parameter to consider during the analysis as it drastically controls the fate of trapping mechanisms at high injection rates.

1. Introduction

CO_2 behaves as a supercritical fluid in storage mediums located at a depth of greater than 800 meters [1]. Aquifers located at 1-3 km below the surface are permeable geologic layers saturated with saline water, and confined with a caprock [2]. Due to its geological setting and high storage capacity, aquifer is probably one the best places for CO_2 storage [2, 3].

The major concern of Carbon Capture and Sequestration (CCS) technology is to ensure that CO_2 remains confined [4, 5] when efficient trapping mechanisms immobilize it in the storage site. Trapping of injected CO_2 into a saline aquifer, however, is done through a series of events including immobilization of CO_2 by structural features or capillary forces, dissolution, and chemical reactions of CO_2 with brine compositions and rock minerals [6]. These mechanisms are classified as structural [3, 7-9], residual [2, 8, 10], dissolution [8, 11, 12] or mineral [7-9] trappings which may take place in a saline aquifer medium. The relative contribution of these mechanisms, however, depends mainly on a number of parameters. For instance, residual trapping is a function of pore fluids and rock properties as well as injection rate, and reservoir conditions [13]. Dissolution trapping is sensitive to variation of pH and concentration of different ions in pore fluid [14]. The fate of the mineral trapping, on the other hand, depends mainly on the pressure, temperature, pH, geochemical conditions, and activity of the cations dissolved in water [12]. There are, however, few studies carried out in recent years discussing the impact of these crucial parameters on different trapping mechanisms [15-19].

The aim of this paper is to evaluate the effect of injection rate and concentration of $CaCO_3$ on the structural, residual and dissolution trapping mechanisms of heterogeneous aquifer. A numerical simulation is run through Eclipse software for the purpose of this study and efficiency of trapping mechanisms are evaluated and compared as a function of injection rate and $CaCO_3$ concentration.

2. Simulation approach

In this study, CO_2 storage in an aquifer is exercised using Eclipse Reservoir Simulator to explore the interaction of structural, residual and dissolution trappings in the storage site. It is assumed that dissolution occurs instantly and gas-rich CO_2 is in equilibrium with water under isothermal conditions of 90°C.

The dimension of the aquifer was 180 m × 180 m with a thickness of 15 m. The rock properties (i.e., porosity and permeability) of five layers of fluvial sand and shale were considered in the model. Figure 1 shows the heterogeneity level of the model in term of horizontal permeability. The model was brought to static equilibrium by considering an initial aquifer pressure of 300 bar at a depth of 2500 m. The model was 100% water saturated without any dissolved gas at the beginning of the simulation. The three prime directions of the model were uniformly meshed by 19 cells in the x-direction, 28 cells in the y-direction and 5 cells in the z-direction. The data set of input parameters used for static modeling of aquifer is listed in table 1.

An injection well was taken into account in the center of the aquifer at a depth of 2400 meter, where only resident brine and CO_2 are interacting in pore spaces. Pure CO_2 was injected for 35 years into the aquifer which appears as a supercritical fluid under injection conditions. No shut-in period was set in the simulation for creation of a driving force due to differential density. Along CO_2 and H_2O, the salt components NaCl, $CaCl_2$ and $CaCO_3$ were considered which can be present in both the aqueous phase and the solid phase during precipitation process. Different schemes were analyzed, aiming to find the effect of flow rate and $CaCO_3$ concentration on the trapping mechanisms. Table 2 presents an overview of simulation cases included in the numerical simulation. H_2O with 0.90 mole fractions and CO_2 with 0 mole fraction were adapted for three cases as given in table 3.

Fluid properties such as gas and brine densities (effect of salt and CO_2) were determined through utilization of equations of state and Ezrokhi's methods [20]. Molecular diffusion was introduced at the beginning of injection for creation of diffusive flow in terms of liquid mole fraction between grids and minimizing the interaction with the formation water. Relative permeability data of Viking sandstone was used for modeling purposes. Figure 2 displays the relative permeability functions considered through the simulation.

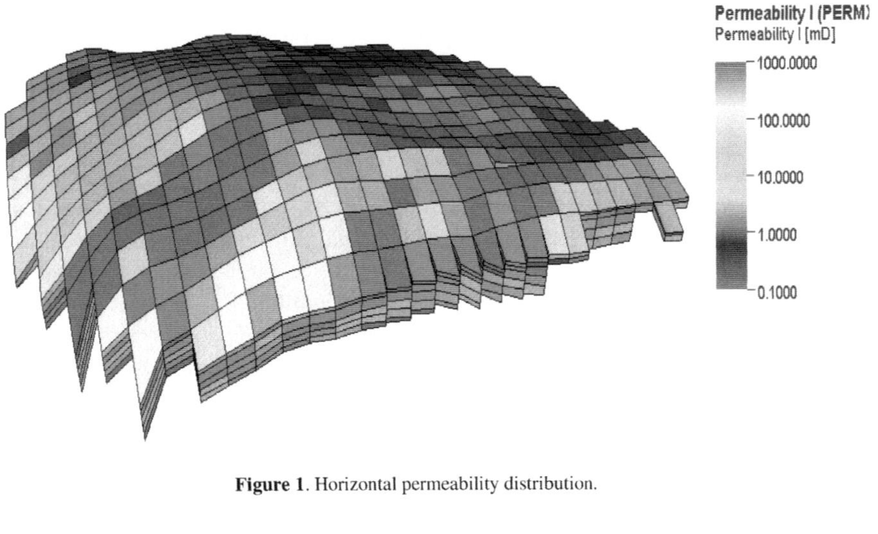

Figure 1. Horizontal permeability distribution.

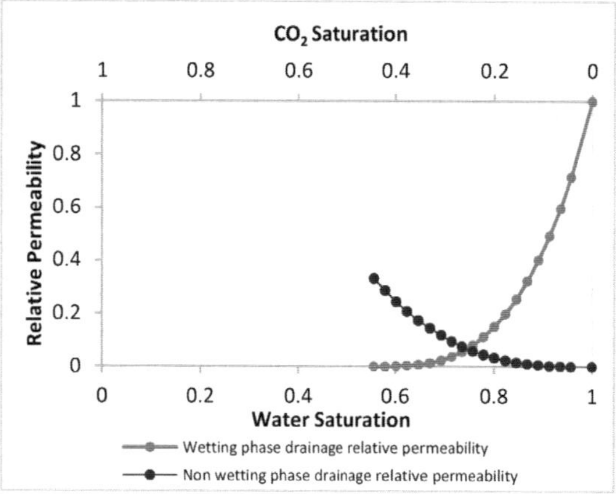

Figure 2. Relative permeability function of CO_2-H_2O system of Viking Sandstone.

Table 1. Basic input parameters for static modeling.

Parameter	Value	Parameter	Value
Irreducible water saturation	0.558	Average porosity	20%
Residual gas saturation	0.442	Average permeability	100 mD
Grid blocks	2660	Top depth	840 meter
Active cells	1760	Average reservoir thickness	15 meter
Length	180 meter	Initial pressure	300 bar
Width	180 meter	Initial temperature	90°C
Equilibrium Depth	2400 meter	Bottom hole pressure	400 bar
Overburden gradient	0.22 bar/meter	Phases	CO_2, H_2O & solid
Salt Precipitation	ON		

Table 2. Sensitivity cases and associated parameters.

Level	CASE A		CASE B		CASE C	
	Flow rate	$CaCO_3$	Flow rate	$CaCO_3$	Flow rate	$CaCO_3$
low	3×10^3	0.00	3×10^3	0.01	3×10^4	0.00
low	3×10^4	0.00	3×10^4	0.01	3×10^4	0.01
medium	3×10^5	0.00	3×10^5	0.01	3×10^4	0.02
high	3×10^6	0.00	3×10^6	0.01	3×10^7	0.00
high	3×10^7	0.00	3×10^7	0.01	3×10^7	0.01
high	3×10^8	0.00	3×10^8	0.01	3×10^7	0.02

[a]Components concentration in mole fraction

[b]Flow rate in sm^3/day

Table 3. Initial concentration of different components (0.90 H_2O and 0 CO_2).

CASE A			CASE B			CASE C		
NaCl	$CaCl_2$	$CaCO_3$	NaCl	$CaCl_2$	$CaCO_3$	NaCl	$CaCl_2$	$CaCO_3$
0.075	0.025	0.00	0.075	0.015	0.01	0.075	0.025	0.00
0.075	0.025	0.00	0.075	0.015	0.01	0.075	0.015	0.01
0.075	0.025	0.00	0.075	0.015	0.01	0.070	0.010	0.02
0.075	0.025	0.00	0.075	0.015	0.01	0.075	0.025	0.00
0.075	0.025	0.00	0.075	0.015	0.01	0.075	0.015	0.01
0.075	0.025	0.00	0.075	0.015	0.01	0.070	0.010	0.02

3. Results

The results obtained are presented in terms of amount and speed of free, residual and dissolved gases as a function of flow rate and $CaCO_3$ in three cases (i.e., A, B and C).

3.1 Effect of flow rate (Case A)

In this section, the effect of flow rate on trapping mechanisms in absence of $CaCO_3$ is discussed. The results of simulations for Case A are plotted against time at different flow rates as ploted in figure 3. As it is seen in this figure, increase of flow rate changes the amount of free CO_2 (structural trapping). It is noted that there is a linear relationship between the amount of free gas and flow rate for up to 10 to 20 years depending on injection rates. This relationship, howevere, starts to decline after 20 years due to rapid and continuous pressure buildup. Figure 4 shows the relationship between injection rate and immobile CO_2 in the model.

Showing in figure 4, there is a linear relationship between injection rate and immobile CO_2 (residual trapping) which is aligned with the findings of studies presented in the recent years [17-19]. This relationship is, however, significant only at medium and high injection rates (i.e., 3×10^5 and 3×10^6). In terms of residual trapping though, a similar trend is observed at high injection rates which increases rapidly up to 10 years and then becomes almost constant. This indicates that handling the flow rate would be useful to optimize the capillary trapping.

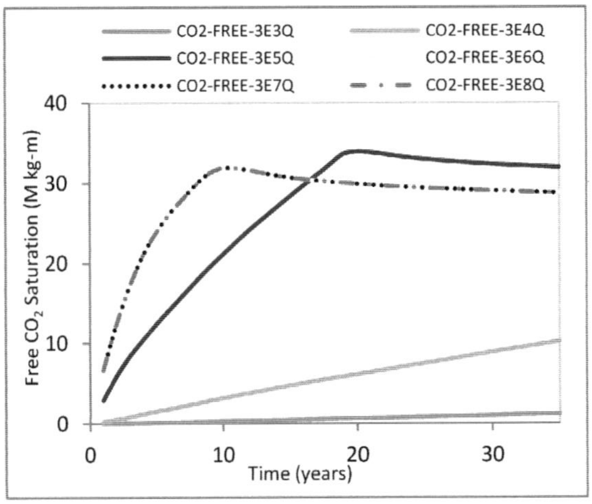

Figure 3. Effect of injection rate on free CO_2 saturation (Case A).

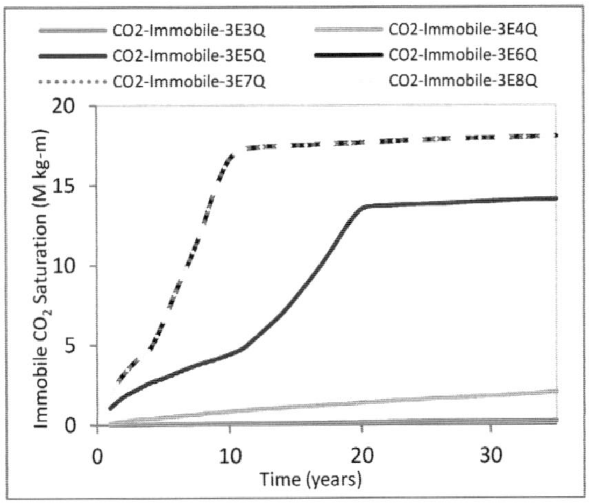

Figure 4. Effect of injection rate on immobile CO_2 saturation (Case A).

Variation of injection rates on the amount of dissolved gas (dissolution trapping) revealed a very same trend for free and immobile CO_2, as shown in figure 5. In fact, there is a rapid increase in CO_2 dissolution at medium and high injection rates immediately after injection begins, which is probably linked to dissolution kinetic factors [8]. The increase in solubility of CO_2 during continuous injection can also be attributed to pressure build up during the injection [21]. It is worth mentioning that the difference in the amount of dissolved CO_2 becomes smaller at medium and high injection rates. Figure 6 compares the trends obtained between trapping mechanisms and low injection rate (i.e., 3×10^3).

The results obtained from such comparison indicated that reservoir pressure builds up linearly during the injection period and structural trapping would be dominant at that stage. This means that low injection rate is not favorable for residual trapping and it may increase the risk of losing cap rock integrity due to long term exposure to free CO_2. A similar conclusion was made by Zhang and Song. (2014) as they indicated that structural and stratigraphic trappings are significant at low injection rate. Figure 7 compares the trends obtained between trapping mechanisms and high injection rate (i.e., 3×10^7).

The results obtained from curves included in figure 7 indicate that dissolved trapping is much significant than residual trapping. It can be concluded that high injection rate favors dissolution kinetic factor and suppress the capillary forces. It is important to note that free gas starts to decline after 10 years onwards at high injection rate when reservoir pressure becomes constant.

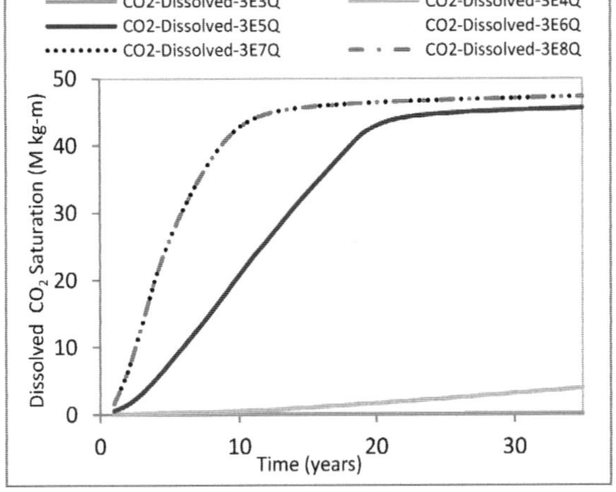

Figure 5. Effect of injection rate on dissolved CO_2 saturation (Case A).

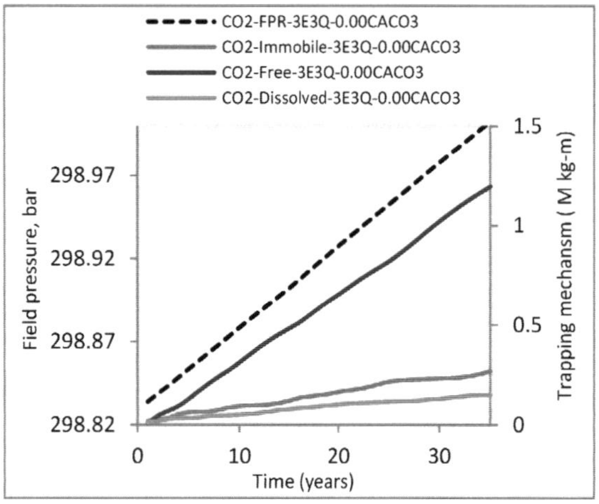

Figure 6. Comparison of trapping mechanisms at lower injection rate (Case A).

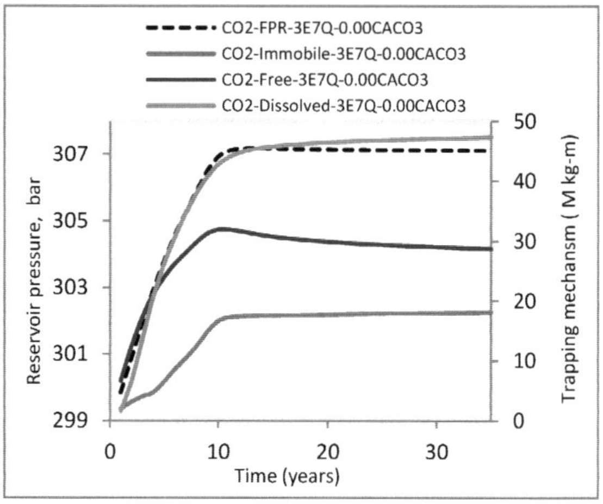

Figure 7. Comparison of trapping mechanisms at high injection rate (Case A).

3.2 Effect of flow rate in the presence of $CaCO_3$ (Case B)

In this section, the effect of flow rate on trapping mechanisms in presence of $CaCO_3$ is discussed. The results obtained for case B are presented in figures 8 to 10. No major impact is observed at low injection rates while trapping at high injection rates is significant. A very same analysis is done for Case B to evaluate the trends of free, residual and dissolution trapping mechanisms at low and high injection rates. The results obtained indicate that presence of $CaCO_3$ component in sandstone is offering more CO_2 entrapment during injection period (35 years) at medium and high injection rates as summarized in table 4.

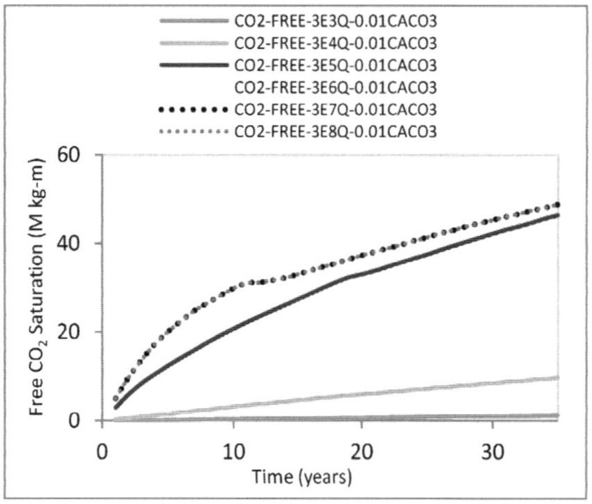

Figure 8. Effect of injection rate on free CO_2 saturation (Case B).

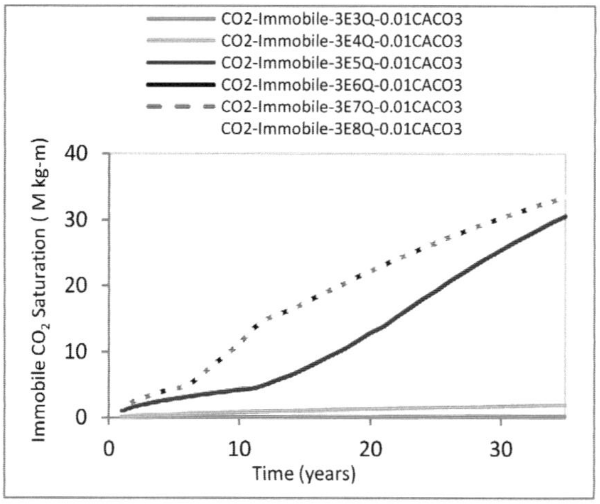

Figure 9. Effect of injection rate on immobile CO_2 saturation (Case B).

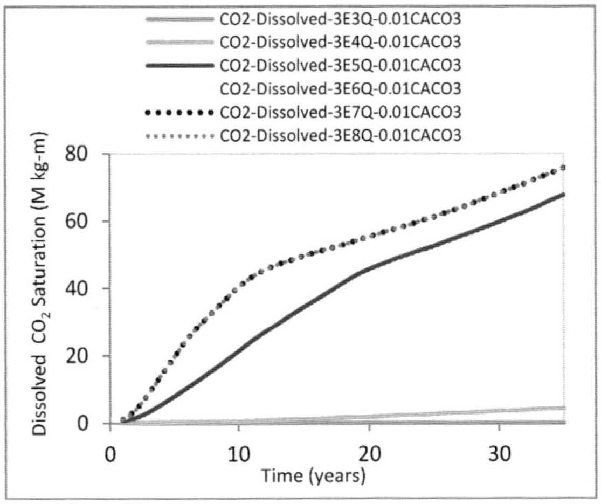

Figure 10. Effect of injection rate on dissolved CO_2 saturation (Case B).

Table 4. Comparison of trapping mechanisms efficiency with and without $CaCO_3$ component effect.

Flow rate, sm³/D	Structural Trapping (M kg-m)		Residual Trapping (M kg-m)		Dissolution Trapping (M kg-m)	
	0.00 $CaCO_3$	0.01 $CaCO_3$	0.00 $CaCO_3$	0.01 $CaCO_3$	0.00 $CaCO_3$	0.01 $CaCO_3$
3×10^3	1.20	1.10	0.26	0.24	0.16	0.15
3×10^4	10.20	9.70	2.00	2.00	4.41	3.88
3×10^5	31.90	46.40	14.10	30.50	45.60	67.50
3×10^6	28.70	48.80	18.00	33.50	47.34	75.70
3×10^7	28.70	48.80	18.00	33.50	47.34	75.70
3×10^8	28.70	48.80	18.00	33.50	47.34	75.70

*Concentration is given in mole fraction

3.3 Effect of $CaCO_3$ (Case C)

In the last part of the study, the effect of $CaCO_3$ concentration on the trapping mechanisms is evaluated and discussed. Three $CaCO_3$ concentrations (i.e., 0.0 0, 0.01 and 0.02) are considered for the purpose of this study to evaluate their impacts on the trapping mechanisms. The results obtained are shown in figures 11 to 13.

The trend in figure 11 shows that there is no significant effect of $CaCO_3$ concentration on the free gas amount (structural trapping) at low injection rate (3×10^4). On the contrary, this impact is remarkable at high injection rate (3×10^7). Similar observation is made for the effect of $CaCO_3$ concentration on the amount of immobile (residual trapping) and dissolved CO_2 (dissolution trapping) as showing in figures 12 and 13. It is generally concluded that these relationships are not straightforward at early stages but will become directly proportional during the injection period. This might be due to time-dependent effect induced by $CaCO_3$.

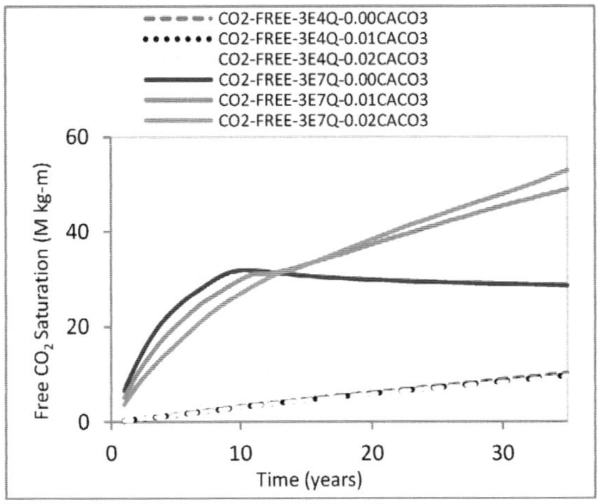

Figure 11. Comparison for the effect of $CACO_3$ component on free CO_2 saturation at 3E4 and 3E7 injection rates (Case C).

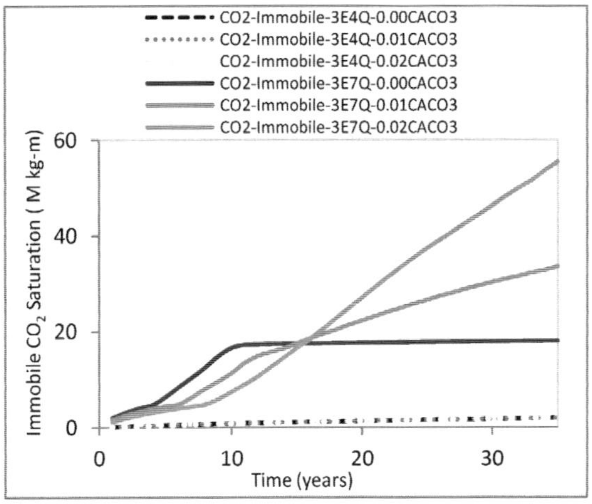

Figure 12. Comparison for the effect of $CACO_3$ component on immobile CO_2 saturation at 3E4 and 3E7 injection rates (Case C).

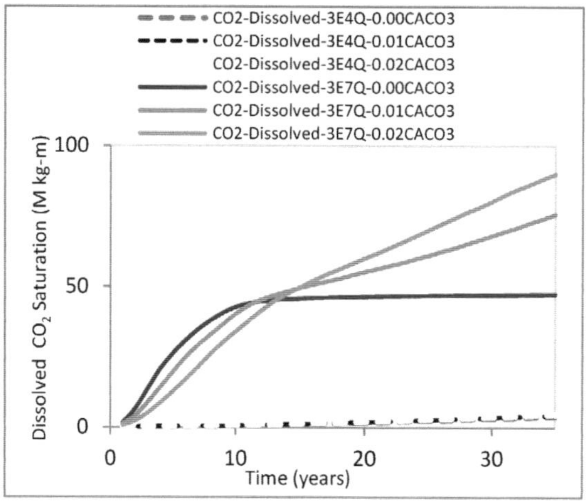

Figure 13. Comparison for the effect of $CACO_3$ component on dissolved CO_2 saturation at 3E4 and 3E7 injection rates (Case C).

4. Conclusions

In this study attempts are made to evaluate the effect of injection rate and $CaCO_3$ concentration on trapping mechanisms of heterogeneous aquifers. The results obtained reveal that selection of injection rate is important to get the highest capacity of a storage medium for CO_2 entrapment. This study also reveals that there is a direct relationship between flow rate and trapping mechanism at very low injection rates. The sequence of trapping mechanisms is also dependent on the level of injection rate. The injection rate is directly linked to the trapping mechanisms when $CaCO_3$ concentration is included in the analysis. The presence of $CaCO_3$ doesn't have any significant impact on the trapping at low injection rates but this effect might have been significant at high injection rate.

References

[1]. Akintunde O, Knapp C and Knapp J 2013 Petrophysical characterization of the south georgia rift basin for supercritical CO_2 storage: A preliminary assessment *Environmental Earth Sciences.* **70**(7): p. 2971-2985.

[2]. Zhao B, MacMinn W C and Juanes R 2014 Residual trapping, solubility trapping and capillary pinning complement each other to limit CO_2 migration in deep saline aquifers *Energy Procedia.* **63**: p. 3833-3839.

[3]. Espie A 2005 CO_2 capture and storage: contributing to sustainable world growth *International Petroleum Technology Conference.*

[4]. Iglauer S, Paluszny A, Pentland C H and Blunt M J 2011 Residual CO_2 imaged with x-ray micro-tomography *Geophysical Research Letters.* **38**(21).

[5]. Juanes R, Spiteri E, Orr F and Blunt M 2006 Impact of relative permeability hysteresis on geological CO_2 storage *Water Resources Research.* **42**(12).

[6]. Taku Ide S, Jessen K and Orr Jr F M 2007 Storage of CO_2 in saline aquifers: Effects of gravity, viscous, and capillary forces on amount and timing of trapping *International Journal of Greenhouse Gas Control.* **1**(4): p. 481-491.

[7]. Benson S M and Cole D R 2008 CO_2 sequestration in deep sedimentary formations *Elements.* **4**(5): p. 325-331.

[8]. Zhang D and Song J 2014 Mechanisms for geological carbon sequestration *Procedia IUTAM.* **10**(0): p. 319-327.

[9]. Saeedi A 2012 Experimental study of multiphase flow in porous media during CO_2 geo-sequestration processes *Springer Science & Business Media.*

[10]. Pentland C H, El-Maghraby R, Iglauer S and Blunt M J 2011 Measurements of the capillary trapping of super-critical carbon dioxide in Berea sandstone *Geophysical Research Letters*. **38**(6).
[11]. Iglauer S 2011 *Dissolution Trapping of Carbon Dioxide in Reservoir Formation Brine-A Carbon Storage Mechanism.* In: Mass Transfer (ed.: Nakajima H), ijeka.
[12]. Ketzer J M, Iglesias R and Einloft S 2012 *Reducing Greenhouse Gas Emissions with CO_2 Capture and Geological Storage* in Handbook of Climate Change Mitigation, W.-Y. Chen, et al., Editors, Springer US. p. 1405-1440.
[13]. Raza A, Rezaee R, Bing C H, Gholami R, Hamid M A and Nagarajan R 2016 Carbon dioxide storage in subsurface geologic medium: A review on capillary trapping mechanism *Egyptian Journal of Petroleum (EGYJP)-Elsevier (in press).*
[14]. Solomon S 2006 Criteria for intermediate storage of carbon dioxide in geological formations *The Bellona Foundation*, Oslo.
[15]. Kimbrel E H, Herring A L, Armstrong R T, Lunati I, Bay B K and Wildenschild D 2015 Experimental characterization of nonwetting phase trapping and implications for geologic CO_2 sequestration *International Journal of Greenhouse Gas Control*. V**42**: p. 1-15.
[16]. Blunt M J and Scher H 1995 Pore-level modeling of wetting *Physical Review* E. **52**(6): p. 6387.
[17]. Nguyen V H, Sheppard A P, Knackstedt M A and Pinczewski W V 2006 The effect of displacement rate on imbibition relative permeability and residual saturation *Journal of Petroleum Science and Engineering*. **52**(1): p. 54-70.
[18]. Wildenschild D, Armstrong R T, Herring A L, Young I M and Carey J W 2011 Exploring capillary trapping efficiency as a function of interfacial tension, viscosity, and flow rate *Energy Procedia*. V**4**: **p.** 4945-4952.
[19]. Soroush M, Wessel-Berg D, Torsaeter O and Kleppe J 2013 Investigating impact of flow rate and wettability on residual trapping in CO_2 storage in saline aquifers through relative permeability experiments *Energy and Environment Research*. **3**(2): p.53.
[20]. Schlumberger 2014, Technical Description 2014.2.
[21]. Bahadori A, Vuthaluru H B and Mokhatab S 2009 New correlations predict aqueous solubility and density of carbon dioxide *International Journal of Greenhouse Gas Control*. **3**(4): p. 474-480.

Acknowledgments
This work was funded by Curtin University Sarawak Malaysia through the Curtin Sarawak Research Institute (CSRI) Flagship scheme. We used the relative permeability CO_2-H_2O data available at Alberta Geological Survey (AGS) and static modeling data from Juanes Research Group (JRG), Massachusetts Institute of Technology. Schlumberger Malaysia is also acknowledged for the Petrel tool and Eclipse Reservoir Simulator licenses.